CLASSROOM USE OF A
CLASSROOM RESPONSE SYSTEM

What CLICKERS Can Do

for Your Students

John Broida
University of Southern Maine

PEARSON
Prentice
Hall

Upper Saddle River, NJ 07458

Editor-in-Chief: *Dan Kaveney*
Acquisitions Editor: *Andrew Gilfillan*
Project Manager: *Crissy Dudonis*
Assistant Managing Editor, Supplements: *Karen Bosch*
Production Editor: *Donna Young*
Supplement Cover Manager: *Paul Gourhan*
Supplement Cover Designer: *Christopher Kossa*
Manufacturing Manager: *Alexis Heydt-Long*
Manufacturing Buyer: *Ilene Kahn*
Manager, Cover Visual Research & Permissions: *Karen Sanatar*
Cover Image: *EyeWire Collection/Getty Images/Photodisc*

Printed in the United States of America
10 9 8 7 6 5 4 3 2 1

ISBN 0-13-187900-6

Pearson Education LTD., *London*
Pearson Education Australia PTY. Limited, *Sydney*
Pearson Education Singapore, Pte. Ltd.
Pearson Education North Asia Ltd, *Hong Kong*
Pearson Education Canada, Ltd., *Toronto*
Pearson Educacion de Mexico, S.A. de C.V.
Pearson Education—Japan, *Tokyo*
Pearson Education Malaysia, Pte. Ltd.

PART 1 | The System — 9

1 | Introduction — 13
This Technology is Not Hard To Learn To Use — 14
Time: The Only Real Downside Of The Use Of Clicker Technology — 16
What Can Clickers Do? — 19
Why Use Clickers Rather Than Something Else? — 21
Clickers Are An Extension Of PowerPoint — 24
Clickers Don't Have To Change Everything — 28

2 | System Mechanics — 29
The Clicker — 29
The Receiver — 35
The Program — 37
The Computer — 39
The Projector — 42

3 | Student Reaction To Clickers In The Classroom — 43
Warning Students About The Novelty Of Their Situation — 44
What My Students Say About Clickers — 46
Comparing Clicker And Traditional Classes — 50
Reactions Of Students In Other Classes — 52

4 | Faculty Responsibilities — 55
Make Sure The System Will Work In The Classroom
You Want To Use It In — 55
Use clickers To Provide Feedback — 58
Students Seem To Enjoy Variety — 63
Students Need To Know That Their Response Was Recorded — 65
Students Dislike Being Held Accountable — 67

PART 2 | Using the System — 71

5 | A Basic Introduction To PowerPoint — 73
Using PowerPoint For The First Time — 75
Adding Clickers To PowerPoint — 78
Some Additional Hints — 80

6 | Taking Attendance 83

To Take Attendance Or Not To Take Attendance,
    That Is The Question 85
Setting Up A Class List So That You Can Take Attendance
    With The System 87
Using The Participant List To Take Attendance And More 88
Dealing With Student Anxiety/Distrust About Your Knowing
    Their Responses 90
Cheating With Clickers 92

7 | Improving Student Study Habits 95

Quizzing As An Effective Teaching Strategy 97
Feedback After The Fact 100
Examination By Clicker 102
Entering Grades Into The Grade-Book 103

8 | Promoting Active Learning And Student Interaction 105

Passive Learning Versus Active Learning 106
Preventing Random Clicking 109
Increased Interaction As The Result Of Clicker Use 112
Have The Students Write The Questions 115

9 | Clickers Change How You Spend Your Time 119

Using Clickers Reduces The Amount Of Time Available
    To Present Information In Class 119
More Interaction With Students 123
It Takes Longer To Prepare For Class When Using Clickers 126
Time Spent Reviewing Student Responses 129
Time Spent In Interactions With Students That Reflect
    How They Respond 131
Clickers Change How Students Interact With Instructors 132

10 | Using Clickers To Promote Active Learning 135

Using Clickers To Promote Discussion 135
Asking Related Questions To Promote More Integrated Thinking 137
Have Students Write Or Revise Questions 138
Use Clicker Questions On Exams 140
In Summary 141

FOREWORD

IN MY OPINION, the first problem that needs to be confronted when considering the use of technology is what I, the instructor, want the tool to do for me. Only once you know what you want the tool to do for you should you consider incorporating any technology into your classes. And then you need to find a tool that does what you want technology to do for you. Reading this text may help you determine if clickers (classroom response systems) are right for you. Please, compare what you want to use technology for with what these systems do. Then decide if clickers are right for you.

This guide, prepared by a user who is very comfortable with classroom response systems, indicates what clickers can do, and may help to reduce the time it takes to use clickers effectively and efficiently. Remember that a good background in the use of this system may take a while to develop (consider the time it will take to read or skim this text), but taking the time now may save you time later. Hopefully this text will enable you to make an appropriate decision about the use of technology in your classroom. This guide outlines the probable costs and benefits, so that you can make an informed decision about the use of this system.

If you want to provide web-based materials, such as quizzes, videos, grade books and/or discussions, then the adoption of a course management system may make sense. If you want to provide overhead notes, talking points, or videos to students as they sit in the classroom, then incorporating PowerPoint, or some similar system, may be more appropriate. However, if you want to increase interaction within the classroom, then a classroom response system may be the best option. Table I indicates the various types of technology available to you, and lists the strengths and weaknesses of each. Please, before you go further, decide what it is that you want to do. Then, find a tool that will help you do it. If I can be of help in your decision making or search, please do not hesitate to contact me. I can be reached at Broida@usm.maine.edu or (207) 780-4255.

TABLE 1 | Technology Available To Instructors

TYPE	STRENGTH	WEAKNESS	FUNCTION
Overhead Projector	Ease of use	Preparation of Materials	Project static images
Presentation Software (PowerPoint)	Ease of use, Flexibility	Hard to use well Often misused	Project static and moving images, sounds, etc.
Web-Based Course Admin-istration (WebCT, Blackboard)	Power, Flexibility	Hard to learn to use, Too powerful for most users	Web-Based discussion, quizzing, e-mail, assignment organizer, grading system
CD Based Exercises	Ease of use, Text specific	Hard for faculty to know when students use this tool; Controlled entirely	Activities, Quizzes, Flashcards, Crossword puzzles, Video and sound Clips, Contents depend on text
Publisher Web Sites	Ease of use, Text specific	Hard for faculty to know when students use this tool; Controlled entirely	Activities, Quizzes, Flashcards, Crossword puzzles, Video and sound Clips, Web-Based version of text, Contents depend on text
Clickers (H-ITT, PRS, Qwisdom)	Ease of use, Flexibility, Power, Based on Power-Point	Time it takes to set up (minimal), Time it takes from other in-class activitie	Creates active learning environment, even in large enrollment classes. Polls student knowledge and opinion

The first section of this text, the system, describes the components of the classroom response system and how they work. It also discusses the effects of the system on the classroom, and how people, both faculty and students, react to the use of the technology. It describes effective use of the system, and why the system may be helpful to faculty interested in educating their students. The second section, using the system, describes how the system may be used to promote specific elements of understanding. Readers should feel free to skip any section that they do not think useful, and, as their experience directs, to return to and review those sections that they find may help adjust the use of the system to better fit their goals and needs. In fact, this text is not intended to be read from cover to cover, but rather in sections as the reader sees fit. Check out those topics that are most interesting, and skip those that you find less useful—you have better things to do with your time.

PART 1 | The System

FACULTY MAY BE CONCERNED when their students sit passively in the classroom, taking notes but otherwise not interacting with the material being discussed. They may wonder if most students understand what is being discussed or, in at least a few instances, if students are paying attention. Some instructors may tell a joke, ask for a show of hands in response to a question, or may ask non-rhetorical questions in an effort to get students to engage with, think about and generally do something more than transcribe information. In other words, instructors are often concerned that students are not engaged in active learning when members of the class sit passively taking notes, doodling, or staring off into space. Anderson and Ambruster (1991) suggest that even the best of such efforts to focus the class on the materials at hand will not be very successful at keeping the students attention and/or actively engaging them with the material. Given this information, some instructors are looking for ways to change what they do into a more active exercise in learning. They may know what they want to do, but not how to do it.

Clickers (more correctly, a classroom response system) can be an effective and efficient tool for increasing attention and engaging students in the material at hand, especially in large enrollment and/or lecture classes. Used properly, they tend to break up the process of passively taking notes, increase interaction with the material, and increase attention on the topic at hand while at the same time increasing interaction with the instructor and with other students in the class. Use of the system may transform the traditional lecture, where one person talks to many, into a more interactive environment. With clickers, one can talk to many and many talk to one (the instructor, but also to each other). If this it encouraged and with a little luck, you may also find that that many will be talking to many, which can be hard to manage but certainly creates an active learning environment.

In answering questions, students may learn what they do not know and so focus their studying more effectively. Furthermore, the instructor can get an idea of which concepts need additional review because a significant proportion of the students do not understand them. Clickers can provide a way for the instructor to hear from and interact with every student in the room, instead of just the few who typically respond to those non-rhetorical questions. The system can provide immediate feedback, to the instructor and to the students, about which information is not well understood, how members of the class feel about a question and what information needs more discussion. In other words, clickers may enliven a classroom, change the orientation and direction of the class and promote learning.

Please note the words "can" and "may" in the above paragraph. While technology can help instructors to achieve various goals, using it does not assure that these ideals will be achieved. When it comes to the effectiveness of any technology, how it is used is probably more important than that it is or is not incorporated into the classroom. Many factors, including such things as the size of the class, the personality of the instructor, the students in the class, and student reaction to the instructor and the technology all interact to determine the actual impact of technology in the classroom. This book provides some hints as to how to obtain the effect you are looking for, but there are no guarantees. Instructors need to play with the system, to determine what works for them given their personality and presence in the classroom and for their students. I encourage you to practice with the system before you introduce it into the classroom, so that you become comfortable with the mechanics. Work out the bugs before you present it to your students; you will feel more confident, and they will experience more rapid success if you practice before you actually use the system. Further, when you introduce the system to the class, tell your students that you are learning the system as they are. Tell them that you are interested in their reaction, and appreciate any feedback that they may provide. This will explain any foul-ups and help students to feel that things will work out in the end, even if the initial experience with the system does not go smoothly.

Instructors can continue the idea that this is a learning experience for all involved by asking for feedback once things seem to be going smoothly. For example, once they are familiar with the system, you may want to ask students how to better achieve the goals that compelled you to adopt the system in the first place. You can even use the system to assess their reaction to its use! The idea is that together you may achieve more than if you force this on your students, especially when the technology does not work correctly. I have found that nothing technological works without the occasional glitch, and having the students on your side before hand may make them more accepting of the system despite these occasional problems.

As noted in the forward, in deciding to use technology, you need, first and foremost, to decide what you want the technology to do for you. When you know what you want, you may be able to find a tool that helps you do it. There are many options available to faculty and students, each with a unique set of strengths and weaknesses (See Table 1 in the Forward for a description of the various types of systems out there what each does, and good and bad elements associated with each). Determining what is available and what each option does is often a difficult process made worse because those most likely to talk with you about a tool may teach in places that are very different from where you are, or are trying to sell you something. Getting an unbiased assessment of the strengths and weakness of each system is often hard, if not an impossible task. Selecting a technology is a lot like selecting the text you want to use; there are many options, but you cannot be sure of the strengths and weakness of each until you spend a lot of time and effort exploring the materials in depth.

Once you know why you want to use a specific technology, you can decide what technology to employ. Determine what each of the various available options does and then which is most consistent with what you want to do. Unfortunately, unless you can find someone to tell you about how to use the tool, this step typically requires that you spend the time necessary to learn how to use the technology for yourself. In other words, you may have to spend many hours playing with a tool, only to find that it does not do what you thought it did. Further, because instructors often have more than one thing in mind when they start to explore the technological options available to them and because technology can often be used to meet more than one goal, you may find that the technology you selected helps you to achieve some but not all of the goals that motivated your search in the first place.

There are also problems associated with determining which of the various competing products that effectively do the same thing. Which of these will work best for you? Because each system has strengths and weaknesses, it may be necessary to try several similar products before finding the one that is most appropriate for your situation. This may explain why incorporating technology into the classroom can be a daunting proposition. Just learning about the various types of technology that are available and what they can do can be time consuming, let alone trying to learn how to use one or more of the specific products to determine which is best for your purposes.

It would be nice to know what you can and cannot do with technology before you spend the time and effort necessary to determine if the system will work for you. This text was developed in the hope that reading it may save you time, and facilitate your thinking about the use of clickers. It indicates what can and cannot be done with a classroom response system and provides some hints as to how to do things effectively and efficiently. In part because Prentice Hall makes a pair of clicker systems available to instructors, the text is not product specific; it indicates what systems can do but says very little about how to do it. With two systems to describe, the text would be unwieldy at best, and very confusing. Fortunately, the companies that make the products sold by Prentice Hall (H-ITT, InterWrite PRS and Qwisdom) encourage instructors to call when questions of difficulties arise. Better still, all three are very good at problem solving. Call them if you cannot get something to work as you think it should, my guess is that they will solve the problem in a matter of minutes. This text provides the general ideas developed by someone who uses the system. I cannot answer many technical questions. But they can and will be happy to provide assistance if and when implementation of these ideas is not going smoothly.

It is my experience that, even though many people encourage faculty to incorporate technology into the classroom, few have the expertise necessary to make that conversion as easy and painless as possible. Unfortunately, though the incorporating of technology is encouraged, it is often hard to find people willing and able to help when problems arise. When push comes to shove, first adopters often feel that they are alone, forced to deal with unexpected problems and expected to determine how best to use technology to help students learn without much help from the administration, peers or the publisher. They may not know who else uses the system, or who to turn to as problems arise. Similarly, they may be embarrassed by their lack of expertise. This is a significant deterrent to the use of technology, one that can be rather scary to those just starting to use the tools now available to them. This text reflects my experience with clickers, what works and what doesn't, and how to use the system effectively. In other words, it is may attempt to fill the void created when you adopt technology for the first time. Please feel free to use it as a guide, and to contact me when you find that you cannot answer a question or deal with a problem. I can be reached at Broida@usm.maine.edu or at (207) 780 4255.

1 | Introduction

Technology can be an important addition to the classroom. It can change the way students interact with the instructor and with each other. It can compel studying by those who would otherwise not be bothered, reward the efforts of those who study, and help the instructor reflect on what she or he wants to say in class before class actually begins. It can energize faculty and students, and it can facilitate the processes of both teaching and learning.

This may explain why some, perhaps many members of the faculty feel pressured to use technology. The interesting thing about the use of technology in the classroom is that, while it may have many advantages, there often a downside that is left unsaid. Faculty know, even without being told, that none of the available tools comes without a cost, no matter what the benefits may be. Many have heard of and perhaps witnessed failures, technological melt downs, when whatever was supposed happen did not because of one problem or another. Such experience makes many instructors wary; they know that whatever can go wrong will whenever technology is introduced into the classroom. But rest assured, this text will provide hints about what can go wrong and how to prevent such problems from happening, even as it describes the benefits and downsides associated with the use of a classroom response system.

Many instructors have heard about someone else in their department or college who is using a new technology for the classroom and likes it. Perhaps an administrator has asked them to consider using technology in their classes because it will enhance the reputation of the school or improve student learning. Perhaps a publisher's representative has indicated that there is a new tool available; one that will make it easier for the instructor to teach their class, or will help students learn the material. Or they may have heard about a peer at another institution who successfully uses these tools in their classes. There may also be considerable pressure from students to keep current, and to do what others at the institution are doing. In other words, there is a lot of pressure to incorporate technology into courses, but, as noted in the introduction to this section, there is often little help available to those brave enough to do it.

How do you decide what to do, which technology to use and how to use it? How do you know if a particular system will work for you? What can go wrong? Is there any advice as to how to make things work better, more efficiently, or easily? What can a person do to reduce the time required to get comfortable with and use a technology in the classroom? These questions and many others are common in first time users of technology. Many instructors want the answers to these questions before they adopt a technology and adapt it to fit their requirements. They are understandably concerned about the use of technology, the difficulty associated with getting started, and know that they will, for the most part, have to go it alone.

This text provides answers to many of the questions that confront faculty as they consider incorporating clicker technology into their classes. Written by someone who has used the system for a while, it is designed to address concerns as honestly as possible, answer questions and address problems before they arise. Because I am not trying to sell or convince you of something, this text is as unbiased an appraisal of the system as I can make it. It provides one persons' perspective on the use of the system; what works and what does not. The text also presents some research findings that suggest that my experience is not unique.

However, I do not have all of the answers. The text will not tell you the way to use the system, nor necessarily the best way to use it, in part because people have different goals, and different uses. But it will provide suggestions as to how to use the system should you decide to incorporate it.

This chapter explores the potential advantages of incorporating a classroom response system into your course, but does not ignore the possibility that there are some significant drawbacks associated with their use. Where possible, I have provided hints as to how to minimize the costs and how to maximize the benefits of a system that I sincerely believe can be very helpful to many, if not all, faculty, especially for those who teach large enrollment (more than 20 students) classes.

This Technology is Not Hard to Learn to Use

In comparison to other technologies that I have attempted to employ to help students learn, clickers are relatively easy to master, both for the student and the instructor. For example, course management systems (notably Blackboard and WebCT) will require most faculty to spend hours mastering the system, and some students seem unable to master the required steps to get things done. In comparison, students get the idea of clickers almost without effort. In no time at all they figure out how to select a response, to point the clicker at a receiver, and to make sure their

responses are recorded. Faculty, especially those who are familiar with
PowerPoint, grasp the mechanics of the system in a few minutes. Those
who have never used PowerPoint but are familiar with elements of word
processing may require about 30 minutes to figure out the basics. In fact, it
is my experience that instructors who are not familiar with PowerPoint
often find themselves incorporating additional uses of that program into
their classes once they are used to using it; clickers become a gateway to
other innovative teaching strategies.

In my opinion, ease of use should not be the only reason you decide
to use clickers or any other technology. In fact, this should not be the primary
reason introduce any technology into the classroom. The best reason to incorporate a technology into your teaching is, in my opinion, that it supports
what you are trying to accomplish in the classroom. If you want a course
management system, so that you do not have to deal with a grade book, go
with that system. But if you are looking to increase student interaction within the classroom and to provide immediate feedback to faculty and students
about what students know and what needs further review, a classroom
response system can be very handy.

That said, rapid mastery is certainly a plus when considering which
technology to incorporate into the classroom. Instead of spending time learning to use the system, you can use that time for other things, including the
creation of slides for presentation, reviewing materials, and writing or attending to other responsibilities. In fact, the ease with which clickers can be incorporated is a primary factor that turns people to this technology rather than to
other types of classroom technology. Furthermore, once started, most faculty
find that they enjoy the system. It helps them accomplish things that they
once thought impossible to achieve, especially in large enrollment courses.

What do clickers do? They can, among other things, engage students,
break up the monotony of note taking, take attendance, encourage discussion, provide feedback to the instructor and the students about what is
understood and what needs more review, and encourage students to complete their reading and other assignments on time. Depending on how they
are used, clickers can act to reward effort, reinforce good study skills and
habits, compel students to come to class, engage students in the process of
learning, reenergize instructors and facilitate direct interaction between student and instructor regardless of class size. Yes, a classroom response system
is easy to use, but, more importantly, it does things that can be useful.

Talking with people about what does and does not work and does not,
how they use technology and their experience with it, can be very profitable.
Those who have used a system often know a great deal about the advantages and disadvantages of whatever technology they are using. As they

learn their way through the ins and outs of the system, they develop expertise in using the system, but also in how the system can be used. They try things, some of which work and some of which do not. Talking with experienced users about how they employ the system can be helpful. In doing so, one important question that needs to be answered is, in comparison to other options, how easy is it to use the system? In my opinion, ease of use is a critical element in incorporating technology. If it takes too long to learn to use, or takes a lot of luck or expertise to get the system to work the way you want it to, and/or if using it is hard for students, fagetaboutit.

Fortunately, clicker software and hardware are easy enough to use that few people have real difficulty incorporating it into the classroom. There may be an occasional glitch associated with first time use; a wiring problem perhaps, a bookstore that does not know where to put the clickers so that they are available but will not be stolen, students who cannot purchase the hardware they need because the bookstore does not obtain enough, and/or the clickers may not communicate very well with the receivers. You should expect that things will not go smoothly the first time you use a classroom response system, if only because so many people are involved in classroom set up, making the clickers available to students and so forth. Importantly, for the most part, instructors and students have little problem adapting to the system.

As Murphy says, whatever can go wrong will, and at the earliest possible moment. In other words, you should expect that some things will not go perfectly the first couple of times you use it in a classroom. Knowing this in advance enables you to plan for the unexpected, and deal with problems more effectively than if you assume that the system will always work perfectly. First time users can expect to forget to do something, or that the folks responsible for wiring the system will make a mistake. And those who are not obsessive and/or compulsive in their slide preparation can expect an occasional miscue. Even experienced users have been known to experience a complete and total failure of the system, as when the power goes out, a hard disk crashes, the hardware that enables reception of the signal from the students disappears, or when someone else using the computer removes the necessary software. In these situations, how the instructor handles their predicament more than the failure of the system can make the difference between success and failure. If the instructor can find a way to adapt to whatever goes wrong in a way that does not punish students for the failure of the system, students and faculty alike will come to recognize, if not enjoy, the foibles of technology.

Time: The Only Real Downside Of Clicker Technology

When talking to faculty, administrators and publishers representatives apparently forget that there are some adverse consequences associated with the incorporation of new tools. As in many situations, those who promote an idea often fail to mention the costs associated with changing from one system to another. In part this may reflect their ignorance of what it takes to use the system, or their unwillingness to admit that the new toy (technology) is not perfect. Faculty considering the use of this system need to know from the outset that incorporating a classroom response system into their classroom will not reduce their workload or cut the time required to prepare for class. Nor will it assure that they become better teachers, get higher teacher ratings or cover more material more effectively. In fact, faculty can and should expect that the incorporation of clicker technology into their courses will add to the time that it takes to prepare for class and perhaps reduce the amount of material that can be covered in an hour of classroom time. These can be important considerations, in part because they reflect the major downside of the incorporation of this, or any, technology into the classroom.

For example, you need to consider the time that it takes to become comfortable with the system, as well as the time it takes to use it once you are up and running. Then too, to use a new system effectively and efficiently, instructors need to know the capabilities of the system. Only then can they determine which features they want to use, which they may want to incorporate later as their comfort and skill increases, and which will not be helpful. Adopters of the system do have to make some adjustments to their existing materials to incorporate the new technology. Overly enthusiastic technology proponents seem to forget these steps. Taking the time to do this takes time away from other, potentially more profitable, activities (writing, attending committee meetings, doing research, reading a good book and relaxing come to mind). Thus, despite what those who encourage its use say, there may be considerable disincentives to incorporating technology into the classroom.

TABLE 1.1 | It Will Take Time To:

---

1. Become comfortable using the system; PowerPoint users, 1-2 hours, others 2-4 hours.

2. Create A Slide Show Using Slides Developed by Prentice Hall; 30 - 60 minutes/class

3. Create New Slides; 4 - 10 minutes each

4. Discuss Slides in Class; 1 - 5 minutes each

5. Record Grades/Responses/Attendance based on Clicker Responses; 5 - 45 seconds/student

---

Becoming comfortable with the system is only the first step of several that are involved in the use of clickers in the classroom. Another cost associated with the use of a classroom response system is the time that it takes to create appropriate materials. In other words, the time that it takes to create slides and presentations will tend to increase the time it takes to prepare to teach the course without the use of clickers. You must weigh this cost with the time it takes to become comfortable with the system against the advantages of using the system. The advantages can be very important, including such things as the increased interaction within and outside the classroom, the ability to know who was in class, the ability to find out who needs help and the increased motivation of students to determine if it makes sense to incorporate clickers into your classes. But these enhancements come at the cost of the time required to develop the skills necessary to incorporate clickers and to prepare slides.

Fortunately, Prentice Hall helps to reduce the preparation time by providing clicker-ready PowerPoint slides for many introductory courses. That said, it should also be noted that you will probably want to edit and perhaps reorganize their presentation so that it is appropriate for the way you teach your class and is consistent with your reasons for using the system. For example, if your primary interest in using clickers is to keep students alert, you will probably want more slides and questions that are directly related to what you have just said or are about to say (Who killed Julius Caesar?). In contrast, if you are using clickers primarily to promote discussion, you will probably use fewer slides, and ask questions that are more thought provoking (How should Brutus have approached Marc Anthony after the killing of Julius Caesar?).

The more you use the system, the less time it will take to prepare for class. In part this is because, with practice, it will take less time to develop slides you want to use. Also, you will be building on the slides you developed. However, other factors may come into play that, if left unchecked, may actually increase the time required to use technology as one's experience grows! Specifically, the need to tweak, to enhance the visual and/or auditory elements of a presentation, may become an end in and of itself. Instead of relying on basics, increased skill and comfort with the process may encourage some people to do more. Faculty who come to view the use of technology as an end in and of itself, who attempt to perfect their presentation instead of focusing on technology as a tool to enhance student understanding of material, may spend much more time than necessary on their presentations. Reduced productivity in other areas may be a direct result of the incorporation of clickers into the classroom, unless those who use these systems lose their focus on the students.

Then there is the class time that the system uses. The time required to allow students to answer the questions is time that could be used to cover more material. Similarly, the time spent going over answers to the questions cuts into time for other in-class activities. As a result, adopters of clicker technology often find that they cover less material than they did before they incorporated this technology into their classrooms. Most however do not regret this loss! Why? Because the inability to cover everything is, in many minds, has more than made up for by the ability to keep students actively engaged, to keep their attention, and to focus them on the task at hand. When required to use clickers, students focus on the topic at hand, not on what they are going to do the minute class is over.

What Can Clickers Do?

Clickers provide a mechanism for getting students to answer questions without anyone else in the class knowing how they responded. Anonymity is important because it means that the instructor and the student are the only people to know if and how a student responded. Further, because the instructor cannot immediately determine how each person in the room responded (they learn this later), the student has time to develop an explanation for their incorrect responses. Thus, the student is not concerned that their incorrect answer will be noted by another student, or that they may be singled out for having an opinion which is different from that held by other students in the class. In this respect, students are very different from raising a hand or verbally answering a question. Many students seem to want to keep their ideas to themselves, especially in a large class. Evidence of this comes from classes where students are asked for a show of hands as to who is gay. The number of people who respond in the affirmative to this question is very different from the number that will respond in the affirmative when using clickers, in part because no one needs to know who they are. Clickers allow these shy individuals, and those who have something they would rather not share with the class as a whole, to participate in the class, without anyone other than the instructor knowing what they said.

TABLE 2 | Clickers Can:

1. Engage students by refocusing their attention and compelling class participation.
2. Promote active learning by encouraging students to consider the answer to each question.
3. Allow anonymous responding which is not possible with a show of hands.

TABLE 2 | Clickers Can cont.:

4. Provide immediate feedback with results displayed in seconds.

5. Immediately inform each student if she or he knows the answer to a question.

6. Inform the instructor which students answered correctly which may need additional help.

7. Indicate diversity of opinion and encourage comparison of responses.

8. Quiz students on their reading or other homework to reinforce keeping up with assignments.

9. Take attendance.

10. Grade student understanding by rewarding correct answers.

11. Be easily incorporated because they require little computer expertise.

The system can be used to evaluate what students know, and whether or not they have been keeping up with their homework. For example, some people use clickers to learn what the class as a whole knows, and what they need to work on. The system also allows the instructor to determine what questions an individual student knows the answer to, and the areas in which they are lacking. Some faculty use the system to learn what students know at the start of the term, mid term and end of the semester. In other words, clickers can be used as a platform to give exams and/or pretests. Others use clickers to determine if students have studied assigned readings or other projects, asking questions that students should be able to answer correctly if they did what they were supposed to. Another use of the system is to test retention of something said in class, either days ago or within the last few minutes. Students are often surprised that they did not remember, or are unable to take advantage of, ideas presented in class, which may compel them to do more than just take notes.

But the system can be used for many other things as well. For example, it is possible to use the system to take (and thereby probably increase) attendance, especially in large enrollment classes. Because each clicker has identifying information that it sends along with the response selected by the student, it is possible to identify which clickers were used, and therefore which students were in class on a given day. And it is possible to assess attitudes, to determine how strongly students feel about or agree or disagree with a statement. For example, one can poll the class on attitudes about current politics, current events and/or their favorite TV shows. As you might imagine, use of the system this way is likely to promote discussion and interaction, both student to student and student to instructor. Similarly, students can, when using the system, be forced to confront common misconceptions, to think for themselves while in class (rather than just being a scribe), and to become more alert and involved with the class(they have to put down their pens, pick up the clicker, think about the answer, and respond accordingly).

Responses can be graded or not, at the discretion of the instructor. They can be entirely anonymous, or anonymous only to other members of the class. Responses can be made to yes/no, true/false, or multiple choice type questions. Some of the more expensive systems allow for responses to essay type responses. Students can be allowed to change their answers several times until polling stops, or prevented from doing so. There are a number of features which allow the student to know that their response has been recorded, how long they have to answer the question, and how many others have answered the question. And students can see how the class as a whole responded to the question, though they cannot determine how anyone else in the class actually answered it.

In other words, the system is a versatile tool that can be helpful in energizing a class, converting what might otherwise be a passive learning environment into something much more active and exciting. More importantly, the system has the capability, but it is up to the instructor to make it work effectively and efficiently.

Why Use Clickers Rather Than Some-thing Else?

Given the variety of technology out there for use in a classroom, it may make sense to consider various options before picking one at random. Each type and product has its advantages and disadvantages, things that it does well and things that it cannot do or cannot do efficiently. As noted earlier, it is important to decide what problem(s) you are trying to solve with technology before you decide which product to select. And, once you find what you want done, you may need to look carefully at several options and/or products before selecting one. My advice is that you consider your options carefully before you invest the time required to become comfortable and confident in whatever you chose. That way, when everything does not go quite as well as planned, you will know that you made the best possible decision based on the evidence at hand. This may encourage you when things get difficult.

That said, one reason to consider clickers is that they are among the easiest forms of technology to incorporate into a classroom. Because the system is based around PowerPoint, a program that many faculty are comfortable with, adding them may take little additional effort. And, because PowerPoint is relatively intuitive, especially in comparison to, for example, course management software, those who are computer literate but who have never used PowerPoint will have little difficulty adding clickers to their classroom routine. In comparison to other forms of technology designed for the classroom, this is perhaps the most intuitive approach to incorporating technology into the classroom.

One reason that clickers are more intuitive than other products designed for the classroom is that the developers of the system tested it in the classroom and they market it specifically for use by instructors. In contrast, course management systems are typically marketed to the information technology folks. In other words, clickers are designed to be used by users (people relatively unfamiliar with computer programming), rather than people who support those who use a particular product (experts in computing and instructional technology). Use of the H-ITT, InterWrite PRS, and Qwisdom systems should be easy to master for most people teaching today. In effect, if you can use a basic word processor, you can incorporate the use of most, though not all, features of clickers without much difficulty. People who are not familiar with the use of spreadsheets will find that they need a bit of help to get started, but only if they want to use the system to take attendance. And, once the spreadsheet is set up, these people will not need additional assistance!

In addition to being easy to figure out, clickers have two significant advantages over other technology designed for the classroom. One is that, though the technology is likely to change in the next few years, the materials you develop and use now are likely to work for many years to come. Slides can be reused year after year, because the computer program used to create them does not change much, or in such a way as to render prior work useless. Think about it. PowerPoint is a widely used program. Developers of the software are not likely to want to anger the millions of current users by making changes that eliminate the utility of existing materials. In this regard, the use of clickers is very different from some other proprietary technology available to faculty, which may change dramatically from year to year in ways that make existing materials useless.

From another perspective, clickers change less than other classroom technologies because they rely more on hardware (physical devices), rather than software (a computer program). Once you have the hardware, it is unlikely that anyone will take it away from you, or make changes to computers that will force you to abandon what you are using. Yes, other options may be out there, and there may be "better" systems down the road. Once you have a working system, because you (in conjunction with your students) own the materials, no one can take it away from you. Meaning that you can, if you wish, continue to use the receivers you have with clickers that students can pass on to the next class. In other words, the hardware can be recycled from semester to semester, so that you do not have to change a thing. While it is true that, at some point, what you have may break or not work with a system upgrade, you may be able to use the hardware for years without difficulty. Most people find that they can use what they have for at least a year, more likely two, before they are forced to give it up for something new. And, again, the slides you develop will work with whatever hardware you are required to upgrade too, reducing the difficulty of adapting to new and upgraded hardware and software.

Unlike course management systems, which use commercially available software housed on servers that are typically beyond the control of the faculty using the system, the software that runs the clicker system is located on a specific computer. This machine may be a laptop that moves from room to room, or be housed in a specific classroom for use by all who use that room. This means that, unlike what happens on servers or computers serviced by people employed by the information technology department, faculty may be able to prevent upgrades to the system in mid semester, or at any other time where such updates may prevent the use of existing materials. This helps to empower the instructor/user, and to assure that what worked at the start of the semester will continue to work through the end of that term. In other words, because you have more direct control over the hardware used with clickers than you do with other systems, you may be able to prevent someone from rendering the technology you use completely useless in the middle of the term.

The other advantage that clickers have over other types of technology designed for the classroom is that, in comparison to other technology out there, creating new materials and adapting older ones is easy and does not take much time. Once the user knows how to use and is comfortable with the system, creating new materials for clickers can be done in minutes, in contrast to the time it takes to develop quizzes, exams, or other materials for a course management system. Thus, instructors can quickly modify their materials to meet their changing needs, create new materials as necessary, and not worry about discarding materials that were not especially helpful. Also, I have found that people are willing to share what they have. For example, Prentice Hall provides many slides for use with clicker systems. And, as people develop their own materials for the courses they teach, I have found that many people are willing to share what they have. In other words, you may be able to develop a large library of materials for your course, without much effort.

Interestingly, clicker software and hardware may or may not work with clickers produced by a number of different companies, not just the systems provided by Prentice Hall. Because all of the companies that produce clickers are using the same technology in their systems, the clicker that the student brings to class can, in theory, be supplied by any of a number of companies. Furthermore, though it pays to ask first, the software necessary to receive and sort student responses may be designed to detect the variety of clickers out there, not just the ones that students purchase for your class. This is important because, as clicker use becomes more common, you may find that there are a variety of clicker types in use in your classroom. This is because students who have already purchased a clicker for another class are not likely to want to spend additional money to purchase a separate clicker for your class. In other words, unless the institution restricts clickers to those provided by a single provider, some students will have clickers provided by Prentice Hall, and others may have systems provided by other publishers. These may or may not be compatible—check with your software supplier. That said, because the system typically does not really care about which company provided the clicker, students need not have a separate clicker for each class that incorporates this technology.

This is not to say that clicker hardware and software are not likely to change in the foreseeable future. In fact, changes are on the way. For example, limitations imposed by the use of an infrared signal between the clicker and the receiver can be reduced or eliminated by the use of radio frequency transmitters. Thus, radio frequency clickers and receivers are on the way. And, as users develop new ideas on how the system might be used, software is tweaked to enable new and different uses of the system. However, given the concerns raised by users and the increased sophistication of programmers, the software that has been developed to respond to the clickers is not likely to suddenly become obsolete or stop working. In other words, slides created in PowerPoint are not likely to stop working as newer versions of software and hardware are developed. It is reasonable to assume that, as hardware changes, it will be designed to incorporate existing materials rather than requiring users to start over. This is because the system is designed to permit and in fact encourage, tweaking. Given the numbers of individuals and corporations that use PowerPoint, it is unlikely that Microsoft or the companies that produce clicker hardware and software are likely to abandon it entirely in the foreseeable future.

Clickers Are An Extension Of PowerPoint

Most computer programs that enable the use of clickers take advantage of a program provided by Microsoft called PowerPoint. This program facilitates the presentation of "slides" that contain information to be considered by an audience. PowerPoint slides are typically used to provide a brief overview of what a speaker is or will be talking about. In education this may mean an outline of a lecture, the key points to be discussed, or a list of terms that the instructor will be using along with their definitions. Many students or others in the audience use these slides as the basis for their notes. Unfortunately, the content on the slides is often copied verbatim by students. This is because it is easy to write what they see rather than to think about what is being said. Question slides, for use with clickers, break up the information slides, and encourage the student to think for him or herself.

For those who are unfamiliar with the program, it is often easy to think of PowerPoint as a product that, to some extent, replaces scribbling on the blackboard. Images (slides) created when using this software can then be projected onto a screen at the front of the classroom. Personally, because I cannot plan well enough in advance to know exactly what images I will want students to learn, I have not completely replaced the blackboard with PowerPoint slides. Instead, I put together a few slides of basic topics that I want students to consider, and write things on the board that come to me as I lecture. For example, a question from a student may need a graphical or written response to get the point across. Rather than creating a slide right there in the classroom (which can take some time and confuse many students), I simply use chalk or a whiteboard, depending on what is in the classroom. But, especially when I am planning to present an image, perhaps an illustration of the brain showing where a particular area is, I will create a PowerPoint slide that shows this, in part because the slide is going to present a more correct image than anything I might draw.

Faculty can use PowerPoint to present more than visual images. For example, some use this program to present video and/or audio clips. In the extreme, with branching, timing and other capabilities, this system can effectively be used to take control of what students see and hear for the entire class presentation, leaving the instructor to start, end and occasionally direct the presentation. In other words, this program can take control of the classroom away from the instructor, should she or he want or allow that to happen. Most people do not use it to that extent; they use it to present various materials that the instructor controls rather than leaving everything up to the machine. Because of the ease of use, and diversity of options, PowerPoint provides an excellent medium for the presentation of information.

The images presented in PowerPoint can, thanks to the hardware and software provided by the publisher, include slides that students will respond to using clickers. Assuming that one had the hardware necessary to send and receive student responses, creating these slides without the software might be possible if one were an expert in the use of PowerPoint.

The software makes the process of creating slides that reflect student responses to questions much easier than it would be without these products. Using the software makes it possible, dare I say easy, for even the most novice computer user to create and use these slides.

Importantly, construction of slides that reflect student responses is, for the purposes of the instructor, not significantly different from any others in a PowerPoint presentation. Both are slides, created within PowerPoint or adapted from other software. The rules governing the generation of response and traditional slides are the same, in terms of font size (as big as possible so that everyone can read the words), structure (both typically have a topic at the top of the slide, and details towards the bottom), and in versatility (one can adjust more than a dozen elements of each slide). The software that comes with the clicker is designed to assure that the slides produced for use with clickers is effectively the same as that used to produce information slides. In fact, software provided by Personal Response Systems is designed to allow the instructor to use standard information PowerPoint slides as the basis of their response slides. In contrast, software provided by H-ITT is designed for use with or without PowerPoint slides. In other words, people who use PowerPoint will not have difficulty incorporating clickers because they can use the same PowerPoint they are familiar with to create this different type of slide.

As I see it, there is a real difference in student responsiveness between classes where clickers are used and those where they are not, even when PowerPoint slides are used in the non-clicker class. Though the classroom presentations both use PowerPoint, the effect of response slides associated with the clickers is very different than that of the more typical PowerPoint slide. Instead of simply copying the information on the slide to their notes, students confronted with a response slide must put down their writing tool, pick up their clicker, think about an answer and press the appropriate key on their transmitter. In the words of a psycholinguist, unlike slides which are designed to provide information, clicker slides are designed to obtain information from the student. The effect, on the class as a whole, of breaking up the lecture, making students interact with the material instead of simply record it, and seeing how others respond goes far beyond that. It transforms the passive task of taking notes into an interactive experience. These are the joys of clickers; instructors create their response slides in an easy to use program that many are familiar with (help is available for everyone).

Response slides are typically multiple choice or true/false (yes/no) questions. They can be designed to assess if students understand the material and to provide feedback about the level of understanding individual students have about the topic at hand. Even though, from the point of view of the instructor, they are simply slides that are based on a template provided as part of the software, they provide a different type of information to students and faculty (did they get it right) than the typical PowerPoint slide. This is, I suspect the beauty of the system; one tool is used to provide both content and feedback rather tan separate tools for each task.

Importantly, you need not use all of the features of PowerPoint to use clickers effectively. If you can arrange the slides in the order that you want and figure out how to start the presentation (click on the F5 key or click on slide show, then view show), you may be "good to go." This is because Prentice Hall has created slides for many texts and courses that you can use with little, if any, adaptation. Also, some slide types, notably Likert type responses (How strongly do you agree with the statement...), yes/no and true/false are more easily adapted to meet the needs of a specific lecture than others. Many instructors start the process of material creation with these slides, because they are easier to create than those that require multiple answers (the real problem with multiple choice questions is coming up with a group of alternative, incorrect, answers). In other words, one can start the semester without knowing how to create materials, and by the end of the semester be very proficient in creating materials.

For some courses, Prentice Hall provides standard information slides and clicker slides. Depending on your goals, there may be little reason not to intermix these, though this is certainly an option for those comfortable enough with file editing to do so. And, as their expertise grows, 0.

Given this, it is not surprising that some faculty who have used the clickers have become so comfortable with PowerPoint basics that they felt the need to learn more, if only so they could do more with the system. In fact, this is a common problem with the use of technology in the classroom; instructors get hooked and want to do more. What was once a stretch becomes too easy, and faculty see ways that they might use the system differently. The result can be a lot of time spent tweaking the system. Unfortunately, getting things to look better, adding sound, flash and other such "bells and whistles" may take away from the effectiveness of the presentation. Worse, at least from some perspectives, is that developing and incorporating these tricks takes time away from other important activities.

Perhaps the real drawback to the use of clickers is that they may cause some faculty to spend time tweaking their presentation rather than focusing on the best ways to get the important concepts across. I find that it is easy to get addicted to exploring the possibilities, rather than focusing on the goals of the use of technology. Having mastered parts of a presentation system, faculty may spend hours learning how to do things differently, only to realize that that technology was not the best way to do get a specific concept across.

People often get hooked into one system, forgetting that there may be other ways to do the same thing. My advice, always ask why you want to do something before you do it. It may save you many hours of tweaking, getting the color just right for example, instead of doing other important jobs.

Clickers Don't Have To Change Everything

Asking questions using clickers does not preclude asking questions that you might ask if you were not using clickers. Instructors can, if they wish, ask open ended or other questions (yes/no, how many agree with the statement..., etc.) in class just as they might otherwise. Of course, such questions require a verbal or other non-clicker based response, unless the class is using one of the more sophisticated and expensive classroom response systems that enable wireless transmission of text. Interestingly, the number of students willing to respond to such questions may be larger in classes where clickers are used than in classes where they are not. This is because students may become accustomed to responding to questions using the clickers, and so be more comfortable about verbal responses. Some might refer to this as the "foot in the door" technique; get students to start answering questions with the clickers and they are more likely to answer questions verbally. Another possible explanation is that students develop confidence as the result of the use of clickers. Perhaps they find that they are almost always correct in their clicker responses, or they realize that they answer more questions correctly than they expected, or more than others in the class do. Regardless of the source, the increased self confidence that accompanies the perception that you know what you are talking about may encourage those who might otherwise be silent to join in the discussion.

And the incorporation of clickers into the classroom is not, in and of itself, going to change how students interact with the instructor. For example, if clickers are used exclusively to quiz students at the start of class to determine if they did their homework or not, the effect on the classroom will not be as dynamic as if they are used to test student understanding of concepts before, as, and after they are talked about. How clickers are used determines a lot about how students feel about them (they will dislike the system if it is used in grading their performance, but enjoy it if clickers are used to provide feedback about class attitudes towards hot topics under discussion). And, of course, how they are used has a great effect on the effect of a classroom response system on the interactions between students and peers and between students and faculty. If you like things as they are, but want to increase student responding, you will find that this increase is possible when using clickers. However, unless great care is taken to preserve the old ways, the more typical effect is to change dynamics within the classroom.

2 | System Mechanics

A classroom response system consists of hardware and software that work together to provide, collect and present information about how students respond to questions presented by the instructor. Clickers (transmitters that send information about which of several choices a student selects and identifying information) are only part of the system. Other parts include the receiver that detects the information sent by the clickers, a software program that organizes that information, and a computer that stores that information. The software also makes it possible for the computer to project which clickers have responded and to project a summary of those responses. Knowing a little bit about how each part works isn't necessary to use the system effectively and efficiently. However, it may be helpful in troubleshooting problems as they occur. And it may help the curious understand a bit more about how the system works. I do not use technical language here because I am not a technician. I am a teacher who uses the hardware and software, and who has had some problems with both on occasion. The goal of this chapter is not to enable the reader to understand the technology, but rather to know where to look if and when things are not working as expected.

This chapter focuses on the mechanics of the hardware and software, rather than how to use the system effectively and efficiently. It describes the components of the system and how they interact to create a system that can help students stay alert and to respond to questions posed by the instructor. The chapter, unlike the others in this text, was written for those who like to understand the technology they use, rather than for those who want to know why or how to use it.

The Clicker

The clicker is very similar to a remote control used to control a Television, DVD player or similar electronic device. There are two basic types of these clickers, one that transmits in the infra red and the other that broadcasts a radio frequency signal. A receiver located within the line of sight of the device can detect infrared signals. In contrast, like the Wi-Fi systems that many people use to wirelessly connect their computers to the internet, the radio frequency clickers are not bothered by obstructions between the clicker and the receiver. Figure one indicates what some of the clickers out there look like.

Most systems indicate when a signal has been sent, typically by flashing as light on and off. Depending on the system employed, the key pad may have anywhere from 5-20 buttons; some key pads have buttons that turn the system on and off (these buttons occasionally confuse at least a few students initially, who will leave the system off yet click repeatedly and wonder why their responses are not being recorded), others turn on only when the student presses their response. Other, more sophisticated systems allow the transmission of text responses, and enable students to respond to short answer or essay questions rather than limiting instructors to a multiple choice or true/false question format. These systems may also enable communication from cell phones, personal data assistants (PDA) or computers (wireless or wired). Importantly, such systems are considerably more expensive than systems based exclusively on button responses.

The system records responses and notes which signaling device is used in addition to the response that was selected. It cannot record which student clicks in, but, when programmed to do so, links the name of the student with the number of the clicker. In other words, it responds to the transmitter, not the student. This means that each student must have their own signaling device. If two students use the same clicker, the machine will record only one answer, typically the last one sent. This explains why it is important that each student bring their own clicker to class whenever they are to be used. This also means that instructors need to consider their options when a student forgets to bring their clicker to class. This can be especially important when grades reflect the number of days that a student does not click in or the number of correct responses made by each student.

Because the system records the last response entered, if allowed to do so (an option that the instructor selects), students can change their minds as they reconsider the question in front of them. Some systems allow the instructor to limit the number of times a student can change their mind on a question-by-question basis. This can be helpful when you want students to respond with their first impressions, rather than their considered opinion. However, it can also produce considerable anxiety, as when students are graded on their number of correct responses;. If they earn points based on the number of correct responses, students will be concerned that their answer was incorrect. If they are unable to change what they come to realize is an incorrect response, students may blame the instructor for not allowing selection of the correct answer. Thus, you will want to think carefully about the ramifications of limiting the number of times students can change their answer.

In addition to sending a signal, depending on the manufacturer, clickers may also contain the hardware necessary to receive a signal. When this is included, it allows the student to know that the computer received their response by watching their clicker. This signal typically consists of a light which flashes (often the same one that flashes when a response is sent) when a choice is selected, this is supposed to increase student confidence in an unreliable infrared system (many clicks are not recorded when using infrared technology, as discussed later). However, it is my experience that students focus on the screen, and so often miss the flashing light. Thus, if you have this feature, many students will not take advantage of it unless you repeatedly remind them to check for the flashing light. It also means that you need not be concerned if this feature is not included on your system.

Another system that students can use to confirm that their response was recorded appears on the screen. Depending on the system used, there may be a box with a number that changes color when a signal from a specific clicker has been received, or the screen may show the number of each clicker that has successfully sent a signal. If the instructor has linked the clicker to the student who owns it students will be able to look for a specific number or their name to determine if their attempts to respond were successful. However, if the instructor does not link names to clicker numbers (does not use a class list feature) the number assigned to a given clicker will vary from day to day. In this case, the number assigned to a clicker is determined by the order in which students respond to the first question of the day; the first to respond gets the number 1, the second gets the number 2 and so forth. This will confuse students, who will be unable to know if their response was recorded.

It is important that students are able to determine if their response was received or not, especially when using infrared clickers, because a large number of responses are not recorded. There are many reasons for this, including something or someone blocking the direct line of sight between the clicker and the receiver, a busy system (the system is in the process of recording another response, poor aim (some students wave their clickers wildly, trying to figure out where to point to get their response recorded), the clicker battery may have died, the clicker may be broken, or the distance between the infrared clicker and/or the receiver may be too far from the clicker to detect the signal. In other words, selecting a button and pushing it does not guarantee that a signal will be received. Thus students may need to click several times before their response is recorded. Importantly, most of the time students get their responses recorded by clicking repeatedly. My experience is that someone is not successful in clicking in on only about five percent of the slide, and it is often the same student(s) who have this problem. Importantly, radio frequency clickers do not seem to have this many failures.

When a button is pushed, the clicker transmits information about which key was selected, along with an indication of the "clicker number" (an identification code that can be used to indicate which clicker selected that answer). When the program is set up to allow this information to be retained by the computer, and you have used the software to create a class list, the inclusion of information about which clicker was used enables the system to determine which student responded, and how they responded to the question. This information can be very important, because it enables the instructor to determine which students were in class on a given day, which students answered correctly, and which students need assistance to understand the concepts discussed. The mechanics of setting up the system to record how each student answered is discussed in Chapter 5 of this text.

The people who produce both the hardware and software are aware of the more than occasional disconnect between the clicker and receiver. This is why they are introducing hardware that relies on radio frequency waves rather than an infrared signal. Clickers that use radio frequency technology send shorter messages, which means that the machine is less likely to receive two signals at the same time. Furthermore, these systems do not require that there be an unobstructed line of sight between the clicker and the receiver.

Interestingly, radio frequency technology works through walls. This may create some interesting problems when responses are not linked to a clicker assigned to your class. Consider the possibility that students in adjacent classrooms are using clickers. When responses are anonymous, students may find that their responses are recorded in both classes. Confusion of this sort will only occur when the instructor is not using a class list that identifies which clickers are assigned to students in her or his class. Used in this way, the system does not know which clickers to exclude from the responses gathered; it cannot limit those responses that it records to only those on the class list. In other words, when using radio frequency clickers, it may be important to use a class list to identify who is clicking in, if only to prevent student answers in other classes from being recorded as answers in yours.

The various buttons on the clickers are labeled as 1, 2, 3 etc, or A, B, C ..., depending on the system used. Students select the "correct" answer to a question by pushing the buttons associated with that response. Importantly, the instructor should make sure that the options on the questions displayed on the screen (numbers versus letters) correspond to what appears on the majority of keypads in the room. This will minimize the initial confusion with the system, though students rapidly adapt to clicking

A when the "correct" answer is 1, B for 2, C for 3 and so forth. Similarly, they will rapidly learn that the number 1 refers to the A on their keypad, B to 2 and so forth.

Though it is not directly related to the topic at hand, you should be aware that, because some key pads use numbers and others use letters, Prentice Hall produces Powerpoint slides that have a bulleted list, rather than either numbers or letters. If you want to adjust these slides to reflect the keys on the clickers, you can do so easily using Powerpoint. Select the slide you wish to adjust, highlight the answers, click on "format", click on "bullets and numbering" and then, select the format that you prefer to use (letters or numbers) from the numbering tab.

For wireless systems, students typically pay for the clickers they use. The cost to the student, typically between $6 and $35, depends on the system used and several other factors. For example, some providers reduce the price of the unit to students when a campus adopts their clicker system, when a student purchases a new text (rather than used) at the same time as when they purchase the clicker, or when large numbers of clicker sales are expected (as when the system is used in a large enrollment class). I have noted a significant price drop in the last year or so, reflecting the increasing use of the system and the reduced costs associated with producing the clickers as the numbers produced increases. Based on this and conversations with various clicker manufacturers, I anticipate that the price of the transmitter will continue to fall.

Students typically purchase clickers from the bookstore, though they may be available on line from some manufacturers. They may be packaged with a text or purchased as a stand-alone item. Another method is to provide a discount coupon with the purchase of a new text. For this reason, faculty who elect to use clickers should inform their Prentice Hall representative that they intend to use the system. If Prentice Hall knows that students will be purchasing a text and clicker, they may be able to reduce the price of one or both items to those students who purchase the clicker along with their text. They should also tell the bookstore of their intent to use the system, so that the store can make arrangements for sale of these items separate from the text (for example, to make clickers available to students who purchase used books). On my campus (where few classes use the system to date, though the number continues to grow logarithmically), clickers are typically sold in conjunction with a text. Increasingly however, students purchase one system that can be used in several classes. In other words, they need not have a separate system for each class that uses a classroom response system—one clicker will work in all of them. This means that, though they are now typically bundled together, clickers are going to be purchased separately from the text used in the course. This may create some difficulty for (resistance from) bookstores, because the clickers are small enough that they fit easily into a pocket. One solution, adopted by many stores that sell the clickers separately, is to keep them behind the counter—students must ask for a clicker when they check out.

Once purchased, the clicker may be used for several years and for several classes (assuming the software used by the faculty teaching the courses permits the use of these multiple systems). Fortunately for the students, the technology employed is relatively stable, those students who purchased their clickers last year will probably be able to use them again this year. Thus, the investment in the clicker system may enable the students to use the system in multiple classes and for several years. Importantly, this will not be true when faculty switch from infrared to radio frequency systems. In other words, radio-frequency systems cannot be used in classes that are using infrared receivers, and visa-versa unless both types of receivers are used and this may create some unexpected difficulty for the computer. Thus, unless all of the faculty on a campus agree to go with radio frequency or infrared, and do not change their minds, students may need to hang on to two clickers.

Not all systems in use today rely on wireless technology! Some institutions have wired classrooms, where clickers are not purchased by students but are hard wired within the room. Wired systems cost more initially, if only because of the cost associated with wiring the room, but are less subject to interference from obstructions and multiple responses at the same time. They are also not portable; students can only use the system in rooms equipped for their use. This means that students cannot forget to bring their clicker to class —it stays in the room. Users of this system who use the technology to take attendance typically assign students to a specific seat. This assures that the instructor knows who is and is not using the system. In contrast, wireless systems are relatively portable and do not require that students sit in a specific seat. Students carry their clickers with them to class, and may use the same clicker in multiple classes.

Some hard-wired classrooms do not have enough hardware to allow each student to have their own clicker. Because the initial classroom response systems were very expensive, and because of the costs of running the wires, some classrooms have been set up in which students are forced to share their clickers with others in the area. When this happens, students in close physical proximity work as a group to select the best response. While this arrangement facilitates group processes, it does not assure that each student contributes equally to the response selected. One response is selected by the group, and that is the one recorded by the system. This limitation probably means that it would be difficult to base grades on a group response.

Hard wiring increases the chance that a response will actually be received by the computer. This is because the direct link between the clicker and receiver assures that nothing, other than responses sent at the exact same time will interfere with the signal sent by the student(s). However, the cost of hard wiring a classroom is considerably greater than for wireless systems, which probably explains both the small number of hard wired classrooms and the proliferation of wireless systems. As campuses become more comfortable with wireless transmission, the necessity of hard wiring will diminish.

The Receiver

Like the clickers, the receiver is hardware. This piece is, effectively, a box, typically with a light on it that contains the hardware necessary to detect the signals sent by the clickers and, in some cases, sends a signal acknowledging receipt of that signal. This device attaches to a computer via a USB or serial port and receives the signal sent by the clickers. Figure 2 shows what various receivers look like. When a signal is received, the light on some systems flashes on and off, indicating to students that their response has been recorded. The receiver may also send a signal back to receivers, causing the light on the receiver to blink, to further indicate that a response has been detected from that receiver.

The infrared signals sent by clickers can be received only within a limited range, perhaps 50 feet (depending on the system and classroom conditions), and only when the clicker is actually pointed in the general direction of the receiver. Obstructions effectively eliminate the ability of a receiver to respond to the signal sent by clickers, just as they do for television remotes. This explains why many rooms that are hard-wired for clickers have receivers close to the ceiling, and why instructors using the system in other rooms often find it advantageous to elevate their receiver to the extent possible. The fewer opportunities for disruption of the signal from the clickers, the better. This is in part why there is a push to convert to radio frequency clickers. Little things, like walls and waving arms, do not prevent reception.

Another problem that instructors may encounter with an infrared system is that, when they have large numbers of students (more than 50 or so, depending on the size of the room, the number of obstructions etc.) in a class, the receiver may be busy processing one signal when another is sent. The result will be that only the first signal is recorded, and the second responder will have to click again to have her or his response received. One way to reduce the interference caused by other clickers is to have multiple receivers placed strategically around the room. The computer takes less time to process signals from receivers than receivers do to process signals from clickers, so having multiple receivers linked together is not a problem for most systems. That said, gangs of receivers can inadvertently be wired incorrectly, preventing the system from working. Make certain that the technicians wiring the system set it up so that the output from one receiver is the input to another (and not the reverse). Also, it pays to make sure that all of the outputs end up at the computer rather than somewhere else (trust me, I have seen this done improperly more often than I care to think about). Remember that the output of one receiver goes to the input of another; when set up this way you should not have any problem with gangs of receivers. When set up properly, having multiple receivers presents no problems for either the students or the instructor. Importantly, unless there are more than 1000 or so students in a classroom, radio frequency systems do not require more than one receiver.

One concern about receivers is that they may be stolen. In other words, students, janitors, or others with access to the classroom may take the receivers if they are not carefully secured. This is a problem only on campuses where the receiver system is hard wired within the room. An alternative is to set the system up each time it is used. This is more difficult but not impossible if you require more than one receiver for the class. On my campus it is common for those of us who teach smaller sections to carry the receivers to the classroom each day, and set up the system before class. This may take a few minutes, but assures that the receivers are there when they are needed. In contrast, the large lecture halls are hard wired, faculty using the system there do not concern themselves with the receivers as they are permanently installed well above the reach of students and almost anyone else interested in removing them.

Receivers require power to work. Though not a problem for hard wired systems, finding the necessary connection to a power source can be more difficult than it should be. The typical approach is to attach the receiver(s) to an electrical outlet. This attachment typically involves plugging a transformer into a wall outlet and then into the receiver. The connection to the receiver can be either direct (there is a connection on the box itself) or to a wire that also goes to the receiver. Unfortunately, connecting to an outlet can be problematic, as some classrooms may not have a conveniently located outlet. Some instructors have to carry extension cords with them when they set up their system (and often worry about someone tripping over the wires), if only to assure that they can use the clickers in a room where the location of wall outlets was not well planned.

Other than that, the receivers are more or less foolproof. Because they are hardware, there is not much to worry about, other than that it may break. I am not aware that this has ever happened, but it certainly seems like a remote possibility. When this happens, the easiest thing to do is to obtain another receiver. Contact the company that manufactured your system or your Prentice Hall representative for assistance.

The Program

As noted earlier, the software that actually does the work of collecting student responses, created by H-ITT, InterWrite PRS or Qwisdom, is provided by Prentice Hall when you decide to use clickers. It converts the electronic signals from the receivers into useful information, including which clicker was used and what response was selected. When the instructor ends the data collection period, by pressing the enter key or left clicking on the mouse, the software then compiles the information so that it can be presented to students. It also adds the information to a file that can be stored for future analysis. In other words, the program is responsible for organizing the raw data into information that can immediately be displayed on the screen, indicating how many or what percentage of responders selected each answer. It also creates what amounts to an excel file so that the information about how each clicker (student) responded to each question.

The program you receive depends upon the company supplying the clickers that you use. While you can use InterWrite PRS software with H-ITT clickers (for example), it typically makes sense to use the software designed for the clickers you are using. In fact, the software you use does not matter from the point of view of the clicker user or the instructor, they are more or less interchangeable and work with all systems. (This has to be the case since students in any class may have purchased their clickers for another class.)

Fortunately, the software comes with instructions on how to install it, and how to use it. These instructions, though occasionally about as clear as mud, will be helpful in getting started with the system, and allow you, the user, to set the program up so that it will work on your computer. Because the companies that provide the software tell you how it works and how to use it, these topics are beyond the scope of this text. If you have difficulty installing the software or getting it to work properly, do not hesitate to contact the people who provided the software (H-ITT, InterWrite PRS or Qwisdom). Your Prentice Hall representative will not be able to help you solve such issues.

All of the various programs designed to work with clickers take advantage of PowerPoint, to a greater or lesser degree, and each has its advantages and disadvantages. Specifically, the software provided by H-ITT is best for users who do not know what questions they will ask in advance. If you are going to ask questions extemporaneously rather than on slides, this is the software for you, as it does not require that there be a PowerPoint slide that serves as the basis for the question. Less spontaneous, more organized lecturers may find the system provided by InterWrite PRS more useful. That said, it should be noted that you can type in questions on the fly when using either the InterWrite PRS or Qwisdom system.

One thing to look out for when using the program for the first time is that, because the system uses either a USB or serial port, some other program on the machine may think that it controls that port. If you have other hardware that uses the same port as the receiver, the message sent by the receiver may be interpreted as information for that other program. Referred to as a port conflict, this can prevent you from receiving information from the receiver unless and until that other program or device is turned off. Doing this is easy, but specific to your system. I strongly encourage you to contact someone on your IT staff and suggest that you have a port conflict.

Another thing to become comfortable with, well in advance of the first class, is selecting where the machine will look for the slides you have created, and where it will put the data from the session once class is over. If the machine you use to present slides to the class is different from the one you use to create them (in other words, if you do not have a laptop computer), you are going to probably have to learn how to access information from a portable drive, and to save it on to that or a similar device. Your options, depending on the systems you will be using, include a floppy disk, zip disk, CD or flash (pen) drive). Another solution is to find a way to e-mail the slides to the computer you will be using, and to e-mail the results of the session back to the machine you use more often. Note, the system you use to carry information to the classroom and the system you use to carry information back to your home machine may be the same—you need not have one drive for the slide show and another for the results. One concern, however, is that when using the same medium for both tasks, there must be enough disk space for the files you are carrying back and forth. This is not usually a problem, most media other than a single floppy disk typically have plenty of room on them for materials used and created in a single class. However, problems are more likely to occur when additional information is on the disk you are using. In other words, it may pay to limit the number and size of other files on any disk used to carry information to and from class.

If you have the chance to do so, I encourage you to play with as many of the available systems provided by Prentice Hall as you can. Find the one that works best for you. But, if you are not given a choice, I suspect that you will find that, whatever you are given, you will be more than satisfied with its performance.

One additional piece of advice. If you find that the program does not work for you for whatever reason, do not call Prentice Hall for assistance. Though they want to help, the technical help people will give you is often the phone number of the clicker provider. Don't waste your time calling Prentice Hall when you can call the clicker company directly. To reach help for the H-ITT system call 479-582-2414 or go to http://h-itt.com to find a help manual. InterWrite PRS maintains help centers in Maryland (800-344-4723) and Arizona (800-856-0732), and their web page is http://www.gtcocalcomp.com/interwriteprs.htm. Qwisdom help can be obtained by calling 800-347-3050 and/or by visiting their website http://www.qwizdom.com/.

The Computer

The computer has several jobs to do as the instructor uses clickers. It runs the program that enables the system to record student responses, summarizes the responses, and projects them so that students can see both the questions and their answers. It also stores how each student responded to each question in a format that allows the instructor to study this information later. Once things are set up properly, the machine will do its job flawlessly, unless there is a system crash or it stores those responses so that they can be recalled later, and it determines what is projected to the students. The software that is running on the machine synthesizes the information collected by the receivers, and uses that data to create charts which can be presented to students within seconds after the students have responded to the question. It is important that instructors who use clickers make sure that the computer they will be using can accomplish all of these tasks in a way that assures that things will work. Checking things out before the start of class is often a good idea because computers, like humans, can be very temperamental.

One of the primary tasks for a computer when an instructor elects to use clickers is to run PowerPoint, which is the program that displays the questions and any other slides the instructor selected. While most office and classroom machines have this program at this point, you may want to check this a month or so in advance of the start of the first class, just to be sure. Allow time for the program to be loaded onto the machine by whoever is responsible for doing such things—typically someone with expertise in information technology. Importantly, if you use a machine with a different operating system, for example Mac, than the one that you will be using in the classroom (perhaps one running Windows), you will need to make sure that slides created on your machine can be read by the one in the room.

That said, it is important to know that users of software provided by H-ITT, InterWrite PRS and Qwisdom need not be concerned about using PowerPoint if they do not wish to use that program to present slides. The software provided by H-ITT allows the instructor to use another medium to ask questions (perhaps an overhead projector, or asking them extemporaneously). In contrast, the software provided by InterWrite PRS includes a slide presentation system that does not require the use of Powerpoint. Those who are not familiar with Powerpoint may find these options useful. However, those who are comfortable using Powerpoint in the classroom will find that adding question slides to their existing presentations is easy and seamless.

Regardless of which system you use, you will have to have the software that runs the clickers on the machine you use in the classroom. This means that you may need permission to load the software on the computer in the room where you will be using clickers, if it is not already there. Some institutions protect their classroom computers and will not allow faculty to load a program on the system themselves. In other words, someone from the IT department may have to load the program onto the classroom computer. Check out the policy at the institution, and for the specific computer you will be using, well in advance of the start of the first class, if only to allow plenty of time to get the work done.

While the program is effectively fool proof, things do not always go smoothly when running this software for the first time. The most common problem when it comes to using this software is the need for a free USB or serial port for the receiver(s). Further, that port cannot be expecting input to another program or otherwise have a port conflict. If you have a conflict, talk with someone in your information technology department about how to fix this. Where possible, they should correct the problem, rather than allowing you to rely on your computer skills, as this can occasionally be a rather tricky process. At the very least, have someone who knows about such things available, if and when you attempt to do this yourself.

If it is not already wired, you may also have to connect the computer to the projector responsible for projecting the image that is on the computer to a screen in the classroom. This typically involves another serial port. Interestingly, with this connection in place, you may lose the ability to see what is on the computer screen, except by looking at the screen in the front of the classroom. Again, I encourage you to talk with people who know the room set up and something about information technology. They may be able to show you how to switch between computer screen, class screen, and both at once, just in case. This is yet another reason to familiarize yourself with the equipment in the classroom before the start of the first class.

Unfortunately, not all classrooms have a projector in them. The institution may have a portable projector that the instructor can use; this is certainly an acceptable solution in most cases. If neither of these options are available to you, you may want to reconsider the use of clickers, if only because students will not be able to see how their responses compare with others in the class. To the extent that one of the reasons to use clickers in the classroom is to provide comparative feedback about how each students' responses compare with those of others in the class, use of the system without a projector is going to be frustrating. It can be done, perhaps with an overhead projector to display the questions and having the instructor or someone else report the percentages of each response verbally for example (you may have to do this if and when the light on the projector blows out), but this looses the immediacy and some of the intimacy associated with the use of these devices.

If you are not lucky enough to have a portable computer that you take with you to class, you will need a way to take your prepared slides from wherever you created them to wherever you intend to use them. This may mean a floppy disk, a zip disk, a flash memory "pen" drive, or e-mail. It is also important to set the clicker program so that it saves the data from your session on to whatever device you intend to use. It may take some practice to determine how to access this information and to determine how to save it where it can be transported elsewhere, though most people who are familiar with computers can figure this out for themselves. Remember that, at some institutions, some systems remove any new information from their drives, including the clicker program, slide shows, and results from the presentation, each time they reboot. Thus it is important that information not only be saved on the classroom machine.

The software that makes the clickers work also relies on the computer to put information from the clickers into a file that it can then use to display how the class responds and for subsequent use by the instructor. Interestingly, this process can, at times, interfere with your ability to get the computer to move forward. For example, if you attempt to end the presentation of a slide by left clicking on the mouse at the same time the machine is processing information from one or more clickers, the machine will not end the slide. Being busy with other things, it effectively decides to ignore your click. This is a temporary situation, clicking again (one or several times depending on how busy the system is), when students are not using their clickers, will result in the slide ending and the results being presented. But be careful about rapid clicks of the mouse—too many and you may find yourself presenting the next slide, or perhaps the one after that, instead of looking at the results from the slide you just presented. That said, do not panic if this happens. You can go back by clicking on the up arrow key.

The Projector

The last piece of hardware that is involved with this system is the projector, the bright light, lens, and associated hardware that projects what is on the computer screen to a screen that can be seen by the students. That said, there are some exceptions to this rule. For example, some larger classrooms may have several projectors linked together, such that all project the same image to various screens located throughout the classroom. Similarly, some systems broadcast the computer screen to television monitors located throughout the classroom. Regardless of the actual setup, the role of the projector is the same—to make what is on the computer screen visible to everyone in the class.

Several things can go wrong however, that make it impossible to use clickers in the traditional format. For example, the light bulb may burn out, so that nothing is projected. It is my experience that students laugh when this happens—they know it is not your fault. You can abandon the use of clickers at that point, or try reading the slides to the students. The clickers will still work and you can report the results to the class. This takes much more time than it would if the projector were working, but has the same effect.

Another potential problem is room lighting when using a projector. There are some classrooms on my campus where the use of clickers requires that I turn out all of the lights! Because the lights are located right above the screen where the image projects, it is virtually impossible to see the projected image with the lights on! Turning the lights off presents other problems, as the students have a hard time taking notes. Similarly, south facing classrooms may be problematic because the sun may shine directly on the screen. If either of these problems occur, you may want to play with various backgrounds and color schemes until you find one that works in the room with the lights on, or request that the class be moved to a more projector friendly room.

On a related note, the color schemes may need to be adjusted to reflect the characteristics of the projector (different machines mix red, green and blue light differently). It may be necessary to adjust the color schemes of the presentation to fit the room. If you find that the colors do not work, or that you want to increase the variety on the screen, you can adjust colors by adjusting the "format series." This is easily done by double clicking on the color area you want to change, selecting, "adjust format series" and then selecting the color or color scheme you want to use. See Chapter 5 for details on using PowerPoint.

3 | Student Reaction To Clickers In The Classroom

The introduction of clickers may have a profound effect on interactions within the class. Student-to-teacher interactions are facilitated because students have the opportunity and/or obligation to respond to questions posed by the instructor. This is very different from the typical classroom where a few students tend to monopolize the opportunity to respond to questions asked by the instructor. Furthermore, questions from the instructor may promote student to student interaction, as they argue various points of view. This is especially true for questions where there is no one right answer. It is even possible to increase the amount of teacher-to-student interaction using this system, if the instructor uses it to monitor student comprehension and progress in the course. These changes in interaction are especially obvious in classes that would otherwise require that students listen to and take notes from a lecture. The passive learning, and loss of attention that accompanies it, can effectively disappear in a classroom where students are responding to questions several times each class period.

Unfortunately, the answers to questions like "will the system work" (achieve the goals of the instructor) and "will the students like using a classroom response system" are less certain. How students and faculty react depends more on how the system is used than anything else. Thus, how faculty incorporate the system, how they react when something is not working and how they use the information provided by the slides effectively determines how everyone reacts to the incorporation of this technology into the classroom.

As you might expect, students will tend to react more negatively when forced to use the system, when they have to pay a lot of money to use it, when they are singled out based on how they respond, and when they are graded based on the number of times they answer correctly than if responding is optional and has no consequences. They will react more positively when they are not evaluated based on how many days they click in, how accurately they answer questions, when the questions asked serve as the basis of discussion, and when their responses are praised rather than berated. While the instructor cannot control the cost of the system to students, her or his decisions about how clickers influence grades, singling out individuals for correct or incorrect answers, and so forth can have a considerable impact on student reaction to the student. And, to the extent that student reaction colors faculty reaction, how the system is used may impact how faculty feel about the incorporation of clickers into their classroom. It's all in how you do it.

A classroom response system can increase attention, increase interaction, increase attendance, and increase the number of students who enjoy the class. This is what has happened in my classes and in others where the system is used. But this result is not guaranteed. This chapter describes reaction to the use of clickers in my classes and the responses of students in other classes where student reaction has been reported. Chapter 4 discusses factors that may tend to increase or reduce student and faculty appreciation of clickers in the classroom.

Warning Students About The Novelty Of Their Situation

Students who have never used clickers before may be somewhat surprised and put off by them, at least initially. This may be because, unlike what they may be accustomed to doing, they may be unable to skip classes without the instructor noting their absence. Further, they may feel that they are being forced to respond to questions that they are accustomed to ignoring. In other words, the schema of the classroom in which clickers are used may be very different from one in which they are not incorporated. Thus, students who like novelty and those who like a challenge may tend to like clickers more than those who are more traditional and set in their ways. Students who have found a successful study strategy that does not include attending class, does not include thinking about course related information while in class, and/or who prefer to go through the motions without being noticed by the faculty will, at least initially tend to have difficulty adapting to the demands imposed by a classroom response system. Those who dislike the inability to interact with the instructor, who do not feel challenged by the opportunity to take notes and not much else from the classroom, and those who are looking for confirmation that their ideas are correct will, in contrast, probably enjoy clickers from the outset. Especially in large lecture halls, where many students become accustomed to doing nothing other than taking notes and hiding when the instructor asks a question, the requirement that they participate in class can be somewhat overwhelming initially. Having never done it before, they may be somewhat reticent to try it. Interestingly, my data suggests once they have used clickers a few times, the vast majority of students want to use them again.

Instructors can do a lot to reduce or prevent this type of negative initial apprehension. For example, if the first questions that a student sees are not graded, or if they are about the use of clickers, current events or something that is not course related, students will quickly become comfortable with and enjoy using clickers in the classroom. On the other hand, if the first few questions that they see remind students that they have not done their homework, result in punishment for incorrect responses, or otherwise have aversive consequences, students may quickly be turned off by the incorporation of clickers into the classroom. With this in mind, I encourage instructors to encourage students to play with the hardware before they are actually used in ways that may result in hurt feelings or punishment. For example, one slide that I occasionally begin my first 8:00 AM class of the semester with asks how pretty the sunrise was that morning. The last option on this slide reads "I was not awake enough to see the sunrise this morning." This almost always gets a chuckle, even as most people select that response. The laughter increases when they see that most of their peers selected that response as well, when everyone realizes how few of their peers are awake in time to watch the sun come up.

Other examples of non-threatening questions that an instructor might use in the first lecture that incorporates clickers include "What is the name of this curse" (yes, curse, not course—many students will miss the "typo" and will enjoy the slide when they see the mistake—it makes the instructor more human), and a true false slide that asks if responding with clickers is worthwhile (people who answer no to that question can be asked why they responded to it if responding was not worthwhile). The idea is to reduce the anxiety associated with the first use of clickers in addition to presenting a little humor. I find that humor is a useful tool for breaking up the tension associated with the first use of the system.

Another idea that may help students become accustomed to the system is to initially allow more time to respond than is allowed later in the semester. It may take a while for students to become accustomed to putting down their pens, picking up the clicker, pointing it at a receiver, and responding. This is especially true when, if you have more than one receiver in the room, some receivers work better than others (are more sensitive, are used by fewer students, or are less likely to be blocked by other objects (students clicking in) than others in the room). It will take a few clicks for students to get used to clicking in, and the less stress associated with those clicks, the better.

Finally, be sure to, at least initially, reward effort. Things may not go perfectly the first time an instructor uses clickers. Expect that whatever can go wrong will, and have a plan to deal with it as it does. For example, if you are awarding credit based on participation and you cannot record participation on a given day, award participation credit to everyone that day. Or, if you are grading questions and one question does not grade properly, give everyone credit for the question. The idea is to make sure that people who did it right get credit, even if it means that a few who did it wrong get credit as well.

While you cannot anticipate everything that will go awry, being flexible when problems pop-up will go a long way towards creating a good feeling about the use of clickers. For example, if you use clickers to take attendance, you may want to allow students at least one free pass before their grade suffers as the result of not bringing a clicker to class and perhaps bring an extra clicker to class for those students who forget to bring theirs. Acknowledging mistakes, listening to student complaints and changing your behavior in response to them, and taking advantage of feedback from the use of clickers to respond when students are having trouble grasping a concept will go a long way toward producing a positive reaction to the use of clickers.

What My Students Say About Clickers

According to the literature (D'Inverino, Davis, & White, 2004; Elliott, 2003; Montgomery, 2004; Trees and Jackson, 2003), students seem to like clickers. They like the opportunity to respond to questions anonymously, they appreciate the opportunity to test themselves, and they appreciate the interaction that may result from how they respond. In the hands of someone skilled in the use of technology who is comfortable with a less scripted lecture, they may work very well, as evidenced by responses to questions asked of my students and others who have taken classes where a classroom response system was used.

The reaction in my classes was similar to what others reported. The overwhelming majority of students in my classes like using clickers. Consider the following selected responses on course evaluations of classes where clickers were used.

TABLE 3.1 | Positive Student Responses To The Question, **What did you think of the clickers?**

- *As a shy student who often holds back verbal participation in class, I enjoy the clickers. I think it offers students that don't have the confidence to speak up a boost of self confidence when they respond correctly without any negative consequences.*

- *Good tools for the classroom. I look forward to seeing them in other classes.*

- *Clickers provided instant feedback as to understanding practice quizzes provided extra study help if desired.*

- *The clickers were a big help. I liked being anonymous while participating in class discussions. I enjoyed quizzing myself with the clicker questions; it gave me a great tool to study by.*

- *I really enjoyed them; they worked well breaking up the class and focusing what we should be paying attention to.*

- *I thought it was a great way for the class to interact and communicate what they were learning.*

- *Being a shy person, who doesn't often want to raise her hand and respond, or ask questions, it was a great way for me to demonstrate my understanding and validate that I was learning.*

- *They are fine...I like to see if I'm on the same page as everyone else.*

- *They definitely facilitated the lectures, and had the class more involved.*

- *The clickers are definitely a good tool for discussion as well as to understanding the material.*

- *They are a nice way to get the class interacting without singling out one student, per se. It is also a great way to see where I stand with the rest of class.*

- *I think the clickers are a great idea. They keep us focused and alive during class time. They also allow us to determine whether we are thinking on the same lines as the class and professor. GOOD IDEA-KEEP THEM!*

In addition to these qualitative assessments, I also obtained some qualitative information. Table 3.2 indicates how students responded to closed end questions asked as part of a course evaluation. Not surprisingly, given the literature, it also indicates that students enjoyed the use of clickers.

TABLE 3.2 | Qualitative Responses To Various Questions About The Use Of Clickers

Question: *The clickers serve to break up the monotony of note taking in class.*

N	Median	Mode	Strongly agree A	B	C	Strongly disagree D
22	1.5	1	11	10	1	0
21	1.0	1	14	6	0	1
			58.1%	37.2%	2.3%	2.3%

TABLE 3.2 | Qualitative Responses cont.

Question: *The clickers serve to refocus my attention.*

N	Median	Mode	Strongly agree A	B	C	Strongly disagree D
22	1.0	1	1	9	1	0
2	1.0	1	11	8	0	2
			53.4%	39.5%	2.3%	4.6%

Question: *The clickers allow me to compare that I am thinking with what others are thinking.*

N	Median	Mode	Strongly agree A	B	C	Strongly disagree D
22	1.0	1	13		0	0
21	1.0	1	12	8	0	1
			58.1%	39.5%	0.0%	2.3%

Question: *The clickers serve to tell the instructor when the students are understanding his/her ideas.*

N	Median	Mode	Strongly agree A	B	C	Strongly disagree D
22	1.0	1	13	9	0	0
21	1.0	1	17	3	0	1
			69.8%	27.9%	0.0%	2.3%

Question: *The clickers are a worthwhile tool.*

N	Median	Mode	Strongly agree A	B	C	Strongly disagree D
22	1.0	1	14	7	1	0
21	1.0	1	15	4	0	2
			67.4%	25.6%	4.6%	4.6%

Question: *I like that the instructor uses the student responses as a basis of discussion.*

N	Median	Mode	Strongly agree A	B	C	Strongly disagree D
22	1.0	1	18	4	0	0
21	1.0	1	17	4	0	0
			81.4%	18.6%	0.0%	0.0%

While the overall response was favorable, it is clear from Table 3.2 that at least one student did not react favorably to the incorporation of clickers into the classroom. Qualitative responses also indicated some negative reactions.

TABLE 3.3 | Negative Student Responses To The Question, **What did you think of the clickers?**

- The "clickers" used in class. I forgot mine twice and had five points deducted each time, even though I was in class, prepared, and participated.... very frustrating....

- Don't use the clickers to take attendance; if students are present but without their clickers, it counts as an absence. I find this unfair.

- I hated the clickers; using them made me feel like a trained seal. If you want to take attendance than do it, but don't make show up just to hit a lever and get my food pellet, I mean credit. It made me feel like I didn't need to pay attention as long as I hit my clicker a couple of times.

- Don't take points away for attendance. I could easily have skipped 10 classes but had someone else use my clicker and gotten credit for being there. Instead, I went to class twice without it and though I was prepared with notes, questions, and actively participated in class, I got points deducted from my overall grade.

- I don't really like the fact that we lose 5 points for every missed class in which we do not click in, how ever I do understand the rationale behind it. It didn't apply to me but I didn't feel that people that might have forgotten their clickers should have lost points if they were still in class, because I feel they still played an active role in class. It seems to me that being there physically is more important that possessing a clicker.

- I learn best when I take notes on a topic, and review them. At times the clickers interfered with this, because I would be distracted from something that I felt was important to write in my notes, to respond to a question with the clicker. It would be helpful if the clicker questions were either at the beginning or end of a lecture topic.

- Please use two reception boxes so that those on extreme sides can click in with easier access. Or just a better spot... not sure where.

- Sometimes responses are merely a reflection of rote memorization of responses to previously introduced concepts.

- I have heard that some classes use them for examinations as well...That would be a mistake.

These negative comments seem to focus primarily on their use to take attendance. These comments probably arise because I deduct five points (out of 550) each day a student fails to click in. Using this system, absence and failure to bring a clicker to class are punished equally. Clearly this policy upset some students. Other concerns, about the quality of the questions, difficulties with the system (there was only one receiver in the classroom), and the timing of when responses were required are useful in that they indicate potential problems with the use of clickers themselves.

Some students dislike the interruption caused by the clickers. They are busy taking notes when they must put down their writing instrument and pick up the clicker. They must start thinking for themselves instead of recording what the instructor just said. This change can confuse some students, at least initially. As indicated in the quantitative data, most students like the interruption caused by the use of clickers.

Some students dislike not being asked to think for themselves on all of the questions. Because some questions asked in class came right after I had provided the answer to the slide, no thought was required. The attempt in doing this was to determine if students understood what was just discussed. Obviously, this was a cause of concern to some students, they wanted to think for themselves rather than regurgitate the information just provided to them. On the other hand, many of the students in the class answered these questions incorrectly, indicating that they were either not paying attention to the slide or the material. In other words, I found it helpful to learn when students were and were not paying attention or grasping the material.

Interestingly, when I cornered students who indicated a negative reaction to the use of the classroom response system, all reported that, though they disliked elements of the system, their overall reaction was strongly positive. Many of those who indicated some distrust or dislike of the system wondered why they are not being used in every class (especially those taught by professors who are not especially exciting to listen to, as one student put it to me).

Comparing Clicker and Traditional Classes

Grades in classes where I used clickers were about 10% higher than in classes where I did not use them. Further, when I compared items on exams that were common to all classes, I found that students in the clicker class got about 15% more of those items correct than students in the classes that did not use clickers. In other words, both measures of learning indicated that students did better in classes where they used clickers.

Importantly, because of the way I graded students, higher grades were not directly the result of additional points earned by using clickers. This is because students in the clicker classes could only lose points for non-use, they did not earn additional points when they used the system. In other words, the increase in grades was not due to points earned from the use of clickers, but was despite the loss of points that occurred when individual students did not use them. And, because the questions answered using the clickers were different from those seen by students in both class types on the exam, responses to clicker questions could not have directly caused improvements on the common questions on the exams.

As someone who teaches research methods, I have to add that although there was a difference in grades and the number of correct answers to questions in common, these differences could be due to several factors other than clicker use. For example, because the classes that used the clickers were taught the semester after the classes where I did not use them, differences could be due to the semester in which they were used (fall versus spring). Alternatively, the classes themselves were very different (the clicker classes were taught at 8:00 am and had a larger percentage of senior level students in them, the non-clicker classes at 5:30 PM). Similarly, my experience (I had an additional semester under my belt when I started using the clickers), experimenter bias (I wanted the clickers to work so may have worked harder in those classes), and a host of other factors could have caused these differences. In other words, there were many potential confounding variables because this was not a study designed to determine the effectiveness of clickers per se. Thus, any conclusions as to the true effectiveness of clickers needs to be carefully considered.

That said, one explanation of the effectiveness of clickers may be that they tended to increase attendance. Though I have no formal data to support this claim, because I did not take attendance in the non-clicker classes, I have the impression that more students consistently attended classes where clicker use was required than in the classes where clickers were not used. This is especially true in large enrollment, lecture classes where students often feel that their absence will not be noted. Also, many students told me that they would not have come to class on several occasions except for the fact that I was taking attendance. Interestingly, others report the same effect—attendance is higher when a classroom response system is used to determine who is and who is not in class than when it is not (Russell, 2003). If in fact attendance does increase when clickers are used, this may explain the increased learning and better grades associated with the use of the system, independent of any of the other benefits associated with the use of a classroom performance system.

Thus, though I believe that the use of clickers improved student understanding of course materials, it is not clear that this is in fact the case. Nor can I attribute any effect of clickers to anything more than what I perceive to be an increase in attendance. A colleague and I are in the process of doing a more careful comparison of the effect of clickers, where some students get clicker questions for even numbered chapters, others for odd numbered chapters, and assessing understanding course content.

Reactions Of Students In Other Classes

The results obtained from students in a research methods course taught by the psychology department at the University of Southern Maine are consistent with what has been reported elsewhere for classes in other disciplines (D'Inverino, Davis, & White, 2004; Duncan, 2005; Elliott, 2003; Montgomery, 2004; Trees & Jackson, 2003). The successes and problems experienced by my students mimic what has been reported elsewhere. For example, D'Inverino et al., (2004) reported that they used the clickers to refocus student attention, since they suspected that students could not focus on a topic for more than about 20 minutes. They found that the system was useful in this regards, as did Horowitz (1988), who reported that only about 48% of students are paying attention at any one time when lecture is the primary mode of communication. Importantly, this study found that active engagement of students increased that percentage to 68%, and that the use of a clicker like system increased it further, to 83%.

The use of clickers also allows instructors to learn that students were not grasping some basic concepts. D'Inverino et al., (2004) also found, that it was often necessary to restructure a planned lecture to reflect the gaps in student understanding. Because the clickers allowed the instructors to determine which concepts were being understood, and which were not, they found it important to reemphasize or repeat some points that were discussed, while others could be skipped, because students needed help in some areas and less in others. The effect was to make the lecture more responsive to the needs of the students, at the expense of the instructors plans.

As indicated by my students, students elsewhere and in other disciplines report that the system is easy to use, and reliable (Elliott, 2003). This study also found that the use of a classroom response system encouraged students to attend class, if only because they were going to learn from the experience. However, students did not like the system as much when answers could be linked to students (Elliott, 2003) and, as in my classes, disliked having the system used to take attendance (Russell, 2003).

D'Inverino et al., (2004) also expressed concern that students may react negatively to the introduction of technology in the classroom because it is inconsistent with their expectations of what goes on in a classroom. To the extent that students expect to come to class, take notes, and walk out, the introduction of clickers may create confusion and discomfort because their use changes what goes on in the classroom. My students did not express such concerns, frequently commenting that they enjoyed rethinking what was supposed to go on in a classroom. Many expressed a hope that other classes and instructors would incorporate clicker technology.

Elliott (2003) introduced clickers as a way of learning what students did not understand about material presented in class. On some days no attempt was made to identify students, and the system was not used to take attendance. On other days, students were linked to their responses, so that the instructor could determine who was and was not responding correctly. As might be expected, students preferred when the system did not identify their responses. The attempt was also to foster active learning in a typically passive, note-taking lecture environment. This paper notes that students were able to monitor their own understanding of material, in addition to allowing monitoring by the instructor of the overall level of understanding in the class. Elliott (2003) also reported that clickers stimulated student interest. However, this paper also notes a drawback in using the system - less material can be covered in lecture because of the time required to present and respond to clicker slides.

Interestingly, once they have seen it used, more faculty incorporate classroom response systems into their classroom. For example, Montgomery (2004) reported that the number of classes using this system increased from 3 in the initial semester to 9 in the next. A similar effect has been noted at the University of Southern Maine, where the number of classes using the system quadrupled after one semester and that number will double again next term.

In other words, it is possible to use the system in ways that students like. More importantly, use of a classroom response system can alter the classroom environment dramatically, in ways that are beneficial for all involved. They may however also be a disaster—lack of preparation and/or failure to understand the limitations of the system can only result in the system producing more frustration than usefulness. Inclusion of this software and hardware does not assure success, however that term is defined. It takes more than hardware and software to make the use of a classroom response system successful.

A note of caution, just because they work well for some instructors does not mean that they are guaranteed to perform effectively and efficiently for you or any other individual instructor. As noted above, student reaction to the introduction of clickers into a classroom is probably more a reflection of how they are used than that they are used. Using clickers does not guarantee an increase in attention, that students will attend class, or that understanding and interaction will improve. What matters is how the instructor asks questions and reacts to the responses. If used to provide feedback to the instructor and student, students are probably going to view their use more favorably than if they are used to grade student understanding of core concepts. Similarly, if students are allowed to respond anonymously, they are likely to like the system whereas if students are publicly called upon to justify their responses, the reaction is more likely to be negative, especially for shy students who would rather not tell others about what they think. And, if use of clickers is optional, if their use is not an indication of who is and who is not in class, responses are likely to be more favorable than when nonuse has a negative impact on grades. In other words, as with any technology, there are potential benefits and pitfalls that will influence student reaction to the introduction of this new tool, toy, or gimmick. The next chapter discusses factors that influence those reactions.

Successful incorporation of clickers into a classroom requires consideration of the reasons for including this technology in the course. Before using them, consider what they can do for you, and if you want them to do it or if some other system might be more appropriate. For example, do you want to use them to take attendance, knowing that students dislike this use of the system? Do you want to provide more feedback than "this is the correct answer." Do you want to use the system to understand what students know, or grade them on their understanding of the material? In other words, success with clickers depends on the goals of introducing them.

4 | Faculty Responsibilities

As with any teaching tool, it is the faculty member who must make it work. And, there is no one best system that will work for everyone. As physicists tell us, every action has an equal and opposite reaction. Doing things one way may produce a desired effect, and it may have some additional secondary consequences that are also positive. However, the decisions made about how to employ a system to do one thing may have adverse consequences in other areas. This chapter describes a number of ways to use the system, and their potential effects, both positive and negative.

Similarly, your use of clickers may differ from what others do with them. Depending on your concerns and your preferences, your goals, and therefore your use of the system, may be very different from what others in your department or college do with them. This chapter presents a number of options; you should decide what to do based on the effects of doing things that way, your approach to teaching, and what you think might work differently for you than for others based on your experience, personality, and ability.

Because student, and faculty, reaction to clicker use depends largely on how clickers are used, there are some important considerations that need to be made before you start to use the classroom response system. This chapter outlines what you can expect from clickers when you use them to perform specific tasks. It also provides warnings about what can go wrong.

Make Sure The System Will Work In The Classroom You Want To Use It In

Many people find that things are likely to go wrong or at least in unexpected directions, whenever you do something for the first time. The first chance for things to go awry is on the first day of class, so expect problems from the get go. Better still, don't just expect them; prepare in advance for the disasters that are likely to occur. Assume that things will not go perfectly each time you want to use clickers, and plan for alternatives. You may find that you need these more traditional approaches to asking questions initially, especially if you have not practiced using the system before going to that first class.

Do not make the mistake of assuming that you can go into the classroom and that the system will work flawlessly the first time out. For example, color schemes may not work well given the lighting in the classroom. Thus, different shades may appear to be the same color, some shades will not show up well, and so forth. In addition, contrast may need to be enhanced due to the amount of ambient light in the room. Other things to consider include the possibility that the location of the receiver(s) may need to be adjusted to assure that few students have problems getting their response sent. Most important, I encourage all faculty, no matter how experienced, to practice using PowerPoint and any other technology you are likely to use before you attempt to use it. Along these same lines, I strongly encourage first time users to take a few minutes to practice using the system before they use it in front of a live class. Have friends, students, and perhaps others interested in the technology watch as you present a short slide show (it need not contain many slides, depending on how adventurous you expect to be initially), and get their feedback. Also, you should look at the slides from various places in the room to assure that they are visible regardless of where students sit. Most people will find things that they want to change. Take notes, fix things, and, if you have time, try it again.

An additional note—these same rules apply when you teach a course in a different classroom. What works in one room may not work as well in another room, with different seating arrangements and lighting. Projectors seem to handle color differently, background lighting may alter the ability to see specific colors on the screen, and other room specific conditions may alter the appearance of your slides. If your images are going to be projected over a video system, be sure to use a dark, rather than light, background as this will reduce the hiss coming from the television. I strongly encourage you to practice, to get an idea of what will happen when you use the system, before you actually try it with your students.

Most people report that it takes at least a couple of days before they are truly and entirely comfortable with the system. The clickers send their signals, the receivers detect them and the computer will record the responses, that is all but guaranteed. The problems come from the different interactions that the system fosters. Specifically, student to student interactions may be different from what you expect, in part because students may be more tempted to talk among themselves as you try to say something when you are done with a slide. They will compare answers even as you explain why one answer was correct and another less so. Student to instructor interactions are also different, in that students are more likely to challenge answers; they may see the question differently, and so have another perspective on the answer. Having communicated with the instructor with their clickers, some may be more willing or eager to engage in other discussion (both in and out of class). Furthermore, instructor to student interactions may dramatically change as you learn how much students missed and how much they understand.

Instructor confidence can greatly reduce student unfamiliarity and anxiety associated with the use of a novel and unfamiliar system. If they sense that an instructor is not comfortable with the system, their fear and distrust associated with clickers will be that much greater. In contrast, if the instructor is at least somewhat confident that he or she is doing things correctly or conveys the idea that they know things will go awry but will be fine in the end, student reaction will be more positive. Confident answers and appropriate attempts to help students who are having problems will do a lot to make the system work better for students during the next class.

TABLE 4.1 | Before The First Class, Instructors Should:

1. Check that their color scheme works in the classroom.

2. Check that the receiver(s) work(s) well.

3. Become familiar and reasonably confident that you can use the system.

4. Check that the necessary software is loaded onto the computer in the classroom.

5. Check that there is no "internal" competition for the Serial Port used by the receivers.

An important consideration at some institutions is that the software must be added to the machine in the room where classes using clickers are held. Clickers will not work if the software necessary to create and use clicker slides is not installed on the machine in the classroom where the clickers are used. This is because the same software that creates the slides also collects the data from the clickers. This may create problems at some institutions, where classroom computers are controlled by people concerned about the impact of the software on other programs that are already on the classroom computer. The program necessary to use the clickers may not work with other programs already there. Similarly, some people are concerned about the possibility that any software added to a classroom machine may contain a virus that will wreak havoc on their systems. Be aware of this from the outset, and talk to the people involved as soon as possible, so that you can address their concerns well in advance of when you will actually use the system.

Another concern is that other programs on the classroom computer may compete for the serial or USB port that the program uses. As described in Chapter 2, port conflicts will make it impossible for the computer to receive clicker signals. Be sure that this is not going to be a problem well in advance of the first class where clickers are used.

Use Clickers To Provide Feedback

One good reason to include clickers as part of a classroom routine is to provide feedback about what students know and do not understand. Students can use this information to determine what they need to (re)learn, and the instructor can use this information to determine what she or he needs to teach. But this outcome is not guaranteed simply because someone incorporates the system into their classroom. Unfortunately, it is entirely possible that an inexperienced instructors' use of the system will benefit no one other than the company that sold the units. I must admit that I know of no instructor, and only a few highly unmotivated students, that actually received no benefit from the use the system, but this does not mean that such lack of utility is impossible.

Use of clickers will not automatically assure success. As with anything else, it is what you do with it that counts. Effective use of clickers can increase student understanding of concepts associated with the course, provide feedback to instructors and students as to which concepts have been mastered and which need more work, and keep students focused on the topic at hand. The system itself does not assure this however, it is the responsibility of the instructor to assure that students get something from their purchase and use of clickers and that the clickers add more value than they cost the students. This is because the instructor determines which questions are asked, when they are asked, how they are asked, and how responses are dealt with. These factors, more than the system itself, will determine the success or failure of the system. Simply put, how clickers are used can negate or enhance their effectiveness.

If the slides used (questions asked) indicate what the students know versus what they need to learn, if the results reflect what the students believe or how they feel, then clickers can be useful to all involved. But if the use of the system does not provide this type of information either to the students or to the instructor and/or keep the students focused on the task at hand, use of clickers will be a failure, regardless of how highly clickers are rated by those using them. Unfortunately, there is no one way to assure effectiveness, instructors must play with the system to determine what works, and does not work, for them personally. You may want to ask students occasionally if they find that the system is working for them, if they value the incorporation of clickers into the classroom, and solicit suggestions on how they might be used better. Having used this strategy a number of times, I believe that there is no harm in asking students what they think, and deciding if their ideas work for you.

Consider a situation in which the instructor presents a series of clicker slides at the start of the class, recording student responses, but not indicating which is the correct answer or promoting discussion of student opinions. Used this way, clickers enable students to learn how many people selected each answer but not much else. They are in the dark as to why one might select an answer different from the one they chose and/or why one response was considered correct while others are not. In effect, used this way, clickers are no better or worse than a paper and pencil quiz in terms of creating an active learning environment. There is no discussion, no education or elaboration, nothing but a noting of response. In other words, the instructor in this situation is not using the full potential of the classroom response system.

However, even used this way, clickers are advantageous to the instructor. Unlike paper and pencil quizzes, the technology allows for automatic scoring of answers. This use alone may explain why an instructor might adopt the use of clickers, they save time and effort when it comes to grading in-class quizzes! Used this way, without providing feedback, other potential advantages to the instructor associated with using clickers in the classroom disappear, and students will probably not enjoy the system very much.

Contrast the previous scenario with one in which the instructor, using the same materials as in the previous example, discusses each slide. In this example, he or she takes the time to indicate why selected answers are incorrect, encourages student discussion of why individual responses were selected and perhaps allows some debate. Students in this situation are provided with the opportunity to learn, not just from their mistakes but from the mistakes made by others. Similarly, students in such a class get a feel for the diversity that exists within the class, that their opinions make sense, and (when they respond correctly) that they learned something. It is obvious that faculty who do not discuss the answers to the questions they pose, do not provide feedback are, in effect, encouraging passive interaction, a situation that is not that much different from what occurs when clickers are not incorporated into the classroom. In contrast, when the instructor takes the time to explain answers and to encourage discussion, clicker use will increase active learning and students will find the system more fun and engaging. Not surprisingly, it is this latter situation that clickers were designed to promote, and that this text is designed to encourage.

Consider the possibility that students are asked a question about how much they like the use of clickers. They respond, and eventually see how the class as a whole feels about their inclusion in the curriculum. The instructor, without any additional comment, then goes on to lecture about something else. What did the students get from this? Not much. They are left with some understanding of how their response compares with the reaction of other students in the class, but have no clue about why they were asked that question, or why others may feel differently about the system than they do. Similarly, no opportunity was provided for suggestions about how clickers might be used more effectively. In other words, the slide had little educational value, for the students or for the instructor. As a result, some students will see this slide, and others presented in the same way, as an exercise in futility. Their response does not seem to matter, to others in the class or to the instructor. In other words, feedback is crucial to the effective use of a classroom response system.

Correcting the problems with this type of clicker use is simple. The slide would have been much more useful if, instead of going on to new material, the instructor had spent a few minutes discussing the class reaction. If they are so inclined, getting students who responded negatively and those who responded positively to explain why they responded as they did would be much more beneficial than effectively ignoring the information from the class. It gives the students the opportunity to consider different perspectives. They can rethink what they believe and to realize why their might be some diversity of opinion. Further, information about why students responded as they did could be useful to the instructor as well - learning what students like and dislike about their use of the classroom response system may enable her or him to change how they use it. It is the discussion after the response that makes clickers useful, not the information about how many or who said what.

If the class size is too large for discussion of opinion slides, the instructor is uncomfortable opening the floor for discussion or there is too little time, then the instructor can answer the question directly. By pointing out pros and cons, indicating why students may disagree and so forth, the instructor creates an atmosphere where students feel understood and appreciated. In other words, where students cannot discuss their ideas, the instructor should present them instead, speaking for the students to validate their opinions or ideas while correcting misconceptions when necessary. Using the responses provided by the students, she or he can perhaps attempt to explain what the responses indicate, at least to the instructor. Students may then be encouraged to discuss this later, perhaps on a discussion board, via e-mail or with each other once class is over. This solution is, perhaps, less ideal than encouraging student discussion, but it conveys the message that the instructor is interested in, and recognizes, student responses. To not mention them suggests that the answers provided by the students are perceived as being less than helpful.

This same idea applies to non-opinion slides, where one or more answers are correct. For example, if the slide were to ask what part of the brain was being pictured, the name of the ship on which Darwin carried out his explorations, or the year in which Columbus began his attempt to head east to China, students who answer incorrectly will benefit more from an explanation as to why whatever they selected as correct was not a good choice than they will from hearing only that they were wrong. This is because students may learn what they missed, in addition to what they do not understand and perhaps why they do not understand it when discussion of the answers follows the slide presentation. And, explaining why an answer might be seen as incorrect is useful to those who selected that response and to others who may have guessed the answer correctly. This explains why some instructors go over each selected incorrect response (some answers may not be selected by any students, these can be ignored or discussed as the instructor sees fit), explaining why each was not considered correct.

Alternatively, students who selected a particular response may be encouraged to discuss why they answered as they did, to defend their idea as it were. Having students justify their answers, regardless of whether or not the response was correct, can be very informative, both for the student and the instructor. In explaining their choice, students may present points that the instructor did not consider that may impact the correctness of a selected response. Similarly, the student may see the question differently from what the instructor anticipated, resulting in a different answer as being correct. And, when the instructor understands where the student's thinking is in error, they may be able to correct their problem, not just for that student but for others who think the same way. When slides are discussed, rather than simply presented, students can learn why one answer is better than another and/or what they missed. In other words, when done correctly, clicker responses can be the basis of a discussion, not simply a response to a question.

Getting student feedback is important, because it may suggest that the wording of the question tricked the students into answering incorrectly or offered an alternative perspective on the question, not a lack of understanding. Importantly, when students attempt to explain their answers, the instructor needs to be careful not to humiliate them. Humiliated students are less likely to contribute to the class, or to express their opinions again. It may be helpful to suggest to the students that the idea behind having students defend their selection is that the defense provides feedback to the instructor about what the student is thinking, and where the student needs help. Conveying this in advance may be increase student willingness to disclose their responses publicly, and so benefit the class as a whole. In contrast, attacking the incorrect answer, perhaps as an indication of a lack of preparation or understanding, can only reduce student willingness to do more than click in.

It is my experience that students occasionally misinterpret what a question is asking. Because they think the question is about something different, their responses may seem incorrect. Allowing discussion of this enables the instructor to see how questions can be misinterpreted, and allows students to see that their understanding of the question was very different from what was intended. This can be very helpful when teaching a class where students come from diverse backgrounds, since such differences may be reflected in different understanding of word meaning and sentence structure. In other words, the use of clickers may enable better understanding of the differences between individuals.

I dislike the use of clickers as an instrument of grading. As I see it, quizzing students, or administering exams via the classroom response system, is asking for trouble. First, because how they respond impacts their grade, students will be more anxious about using the system than they might be otherwise. And, to the extent that they answer incorrectly and receive a low(er) grade as a result, students may blame the system rather than their lack of knowledge. I cannot tell you how often students say that I selected, for example, "A" but the machine recorded by response as "B" when "A" was the correct answer. Then too, students may be concerned about a lack of reliability with the system itself. If their responses are not recorded each and every time they click in, students will be anxious that their responses were not recorded. Because obstructions and the system being busy with other responders assure that not every click is recorded, students cannot help but be anxious when they are graded based on their responses. More importantly, it is easy to cheat when everyone is looking at the same question at the same time. Students have been known to rapidly develop a telegraph system to indicate the "correct" answer to others in the class. For example, it is often easy to see the choice of a neighbor, in part because students typically put the clicker in front of them when they respond and do not make any effort to disguise their selection. Thus, grading based on a classroom response system may not make much sense.

Instead of being used as way to evaluate student understanding, clickers can be used as a study aid, a tool that provides feedback about what students know and need to study. The idea here is that students learn that their response was correct, and so they need not focus much effort in this area, or incorrect, and they need to rethink what they think they know. They then have the opportunity to review, presumably focusing their efforts in areas where they learned they did not understand. In other words, the use of clickers can promote mastery of material even if they are not a good tool for evaluation. Because, traditionally, the classroom is more a place to teach than to test (except of course on those occasions when tests are given), the use of clickers to promote mastery rather than as an evaluation tool is, in the end, more consistent with what goes on in a typical class session than simply asking questions and providing answers.

Students Seem To Enjoy Variety

Lecture classes can be boring. Even the most engaging lecturer is going to have off days, or cover topics that are not exciting. And, when instructors who are not especially dynamic (monotonic, bored, and/or not good at public speaking), the effect on the students cannot but reduce attention in the classroom. This may explain some of the positive reaction that students have to the incorporation of clickers into the classroom. When asked to respond to a question, students have to stop listening and start thinking for themselves. This breaks up the lecture into smaller chunks. This tends to dishabituate students, who can then pay more close attention to the continued lecture then they might otherwise. In other words, clickers provide variety, the spice of life.

Faculty can select from a seemingly infinite number of options when using clickers in the classroom. These include the types of questions asked (Likert type, multiple choice, true/false, and yes/no), font size and style, the color schemes employed (the color of the slide, the text on the slide, the different responses and so forth), and the types of information or graphics that appear on the screen (students appreciate cute art on occasion), and how that information appears (different styles are available). While I personally find such things a waste of time and effort, students have repeatedly told me that they enjoy seeing different things on slides— it helps them focus on the slide. I am yet to find a student who indicated that these embellishments are distracting. In fact, I have heard enough praise of variety that I have adjusted my slides to include considerable diversity in these areas and others. If the goal is to bring students back from their daydreams caused by a lecture, then changes in appearance may be helpful - it certainly seems to get their attention.

Variety can be useful in other ways as well. For example, some questions lend themselves to Likert type responses ("Clickers are helpful," with answers that range from strongly agree to strongly disagree), others fit better into some other format. Further, you might want to use a yes/no or true false answer when asking about a dichotomous topic (are you gay, are you pregnant, the mean of these numbers is 5, for example). Similarly, you may want to use a multiple choice question when contrasting concepts (psychology is: a science, an art, a social science, an easy A). Asking different types of questions will tend to keep students on their toes more than if they come to expect one type of question only, in addition to enabling the instructor to better assess some types of understanding than might be the case if only one question type were available. In fact, you may need to ask a variety of different types of questions to fully assess student understanding of some concepts. Because the software does not limit you to a single type of question, use of different types of questions is easy and uncomplicated.

Similarly, you may want to use different color schemes for different types of information or questions, so as to make finding specific questions when you review the slides that much easier. Another reason to adjust the color is that some colors or color combinations may work better in one classroom than another, depending on the ambient light in the room and intensity of the image on the screen. Again, you will need to try various colors in the classroom you teach in, to assure that the selected hues work well. Different colors may also help break up the monotony of looking at a constant screen. The same can be done with font, including style, size, italics, bold and so forth. In other words, there may be real reasons to use specific schemes on slides. The nice thing is that you can use whatever you wish, because changes and adjustments are easy.

There are additional considerations when thinking about the colors to be used on a slide. For example, given lighting in a room, it may better to present light text on a dark background, or dark text on a light background. I recently learned that, when presenting slides that will be seen on a television monitor, it is important to limit the amount of white background, as this increases the "hiss" produced by the television speaker. Thus, color may be an important consideration when preparing slides for presentation to a class.

Importantly, consider adjusting colors and/or font to meet the needs of students who are visually impaired. For example, the most common problem is an inability to distinguish green from red, which can be dealt with by not using these colors to provide different information. Similarly, yellow and blue are hard to distinguish for some, and a few students may not distinguish color at all. My advice is that you use font, rather than color, when using attributes to distinguish types of slides.

Large font is helpful, especially for students who sit a long way from the projected image. Thus, I tend to use as large a font as possible. But, large font limits the number of words that can be presented. My advice is that you start large, and get smaller only if the amount of information on a slide requires it. If a slide that has a large number of words, use smaller font size (keeping the size as big as possible) so as to get all of the necessary information on one slide. But never use a font so small as to make viewing it hard. I suggest that, before the start of the first class, you determine the minimum acceptable font size by preparing a practice slide with different fonts, standing in the back of the room where the slides will be presented, and seeing what does and does not work.

Students Need To Know That Their Response Was Recorded

Classroom response systems are not guaranteed to record the responses of every student each time they select a response. In fact, in their present iteration (relying on an infrared signal) they miss a large percentage of the responses. When there is an obstruction between the clicker and the receiver, or when the system is busy with other responses, student responses will not be recorded. This explains why students typically click several times before their response is recorded and explains why they need to know that their response has been recorded by the system. To make matters worse, it turns out that, if it exists on the system you are using, a flashing light on the clicker is often insufficient to indicate that a response has been registered. This may be because students are focused on the screen or the receiver, not their clickers, as they actually respond to slides (this problem disappears with practice, but it can create confusion initially). In other words, they may overlook the flashing light, and so need some other confirmation that their response has been noted by the machine. Another possibility is that their attention is on the question, not the device. They are so worried that their answer may not be correct that they ignore everything else.

There are several ways to assure students that their response has been counted. One is, of course, to have them look at the light on the receiver—it should flash immediately after they click (the problem here is that someone else may have indicated their response at the same time). Another, more reliable method, is to remind them to watch their clickers to see that the light flashes, if the system being used has this capability. The best, in terms of assurance for the student, is to use a numbered response grid as part of the slide. Usually positioned at the bottom of the slide, the frame adds clicker numbers as students click in. Students rapidly learn to look for their number as evidence that their response has been recorded. Unfortunately, the area devoted to these numbers takes away from the area in which the question can be displayed. You may find that you switch screens periodically so that students can see the question and then determine that their answer was recorded. You may want to take this into account when developing questions.

Clicker responses can be recorded anonymously, or linked to the name of the student who is supposedly using the clicker. Using a class list, instructors can link the number of the clicker received by each student to their name. Doing this means that the instructor can determine how each student responded to each question. Once this is done, the instructor can link responses to names or not, depending on how they elect to use the system on any given day. Importantly, anonymous responding may be beneficial when asking personal questions (are you heterosexual, how old are you, etc.). However, this technique may lead to less concern about correct responding (the instructor does not know who I am, so it does not matter if I answer correctly or not). Thus, the instructor may chose to link student answers to responses for some class sessions and not others.

Some confusion may arise when using the InterWrite PRS and I-HITT systems because the position of a particular student (number) depends on who else has clicked in on that slide. For example, the first person to click in will see their number in the upper left portion of the screen. When the second person clicks in, this number may or may not move to the left. And the third responder may move one or both of the earlier responders to the left, depending on clicker numbers (the lower number always goes to the left). Unfortunately, when using these systems, students may not be able to see that their response has been recorded because the entire screen is devoted to the question and its answers—leaving no room for the presentation of the clicker numbers of students who have responded. Thankfully, these programs come with information about how to deal with this problem.

Unfortunately, creating the link between clicker number and name is a manual process, one that requires instructor time and effort. Students must provide the number on the back of their clicker to the instructor, who must then enter that value in a spread sheet used by the clicker program that indicates name and clicker number. I have my students e-mail me the number that is on the back of their clicker. Others pass around a piece of paper on which students indicate their name and the number on their clicker. Regardless of how the information is obtained, the instructor, or an assistant, must enter the numbers into the spread sheet. Instructions on how to do this are provided with the software. Note, this can be a time consuming process, but it is not very difficult. One word of caution, you should back up the spreadsheet that contains clicker numbers and names, and put it somewhere other than on your hard drive, just in case.

Selecting to use anonymous responding or to link responses to student names for an individual class session is done just before the start of the class. Before presenting the slide show, the instructor must tell the software if she or he wants to link responses to names. The default is not to link responses to names. Instructions for electing to use a class list come with the software. If you elect to do this, be sure to do it before you start the slide show, it cannot be elected once a slideshow has been started. One other note: when using a class list on a computer that stays in the classroom, rather than transporting the computer to and from the room, you will need to bring this class list to the room each day, perhaps on the same disk that you use to carry your questions and responses back and forth to your office computer.

Students Dislike Being Held Accountable

Some students are shy; they do not want their responses to be identified by the instructor. Other students do not want to be required to use clickers. Many dislike the idea that attendance is equated to use of clickers, and no student likes it when failure to click in on a given question or day results in a loss of points. In other words, students want to be able to use clickers when they remember to bring them to class and do not want to be punished when they forget to bring them or do not respond to a question. Students would prefer that they not be held accountable for mistakes, but get full credit when they respond to a question correctly, a scenario that is impossible to create in the classroom. If an instructor were not to require the use of clickers, or hold students accountable in some way for their use of the system, few students, if any, would use the system. Behavior which is not reinforced will tend not to be repeated, to paraphrase a famous psychologist, B. F. Skinner. Thus, there is some conflict between the students' interest and the need to reward the use of the system.

Fortunately, there are many options available to instructors in terms of how to reward use of the classroom response system. Those who are interested in getting responses, but concerned about student anxiety about being graded on their performance may elect to reward use of the system, rather than accuracy in response. In contrast, those interested in using the system to grade students or who are concerned that students should respond correctly may prefer to reward the number of correct responses. Another option is to assign students to groups, and reward groups based on the number of correct answers by group members, or the number of times members of the group use the system (some systems allow instructors to group students and provide statistics on group, in addition to individual performance). This latter option has the advantage of allowing students to pressure their peers to respond, rather than leaving it entirely to the instructor. The problem with it however is that some students will be angered, perhaps because their grade or other incentive is influenced by the actions of (potentially lazy) group members.

Holding students responsible for their answers increases anxiety. Call it evaluation apprehension or test anxiety, but there is a real concern on the part of some students that they will answer a question incorrectly and fail to earn credit as a result. If students are rewarded based on the number of correct responses, you can expect that students will take longer to respond, as they try to discern the correct answer from some hint in the way the question is worded, and more negative reaction (arguments) when responses are not correct. The additional anxiety associated with this type of grading system will reduce the appreciation of the use of clickers in the classroom, since how they are used has an immediate negative effect on student grades whenever a response is deemed incorrect.

On the other hand, if students are not rewarded for correct answers, there may be a tendency not to care how they respond, at least on the part of some students. Random responding is more likely when students are not graded based on which answer they select, perhaps, as Skinner would say, because correct responding is not rewarded. One technique for minimizing this that I have found effective is to, when responses are consistently incorrect, contact the student directly, expressing concern that the individual is not understanding the material. This has the advantage of personalizing the class, and convincing students that you care about their understanding an progress in the course. I found that taking the time to do this quickly and effectively reduced random guessing, once word got out that I was actually looking at how people respond. The drawback to this way of doing things is, of course, the time required to go over individual responding, and then contacting students. An unanticipated advantage to this approach is that several students noted how much I care about each individual student on evaluations of the course!

Basing reward on the number of responses or number of days that students did not click in, rather than the number of correct responses, reduces the pressure on students to respond correctly. But, it increases the pressure to respond, which some students will not like. For example, if the instructor starts each class with a clicker slide, students who arrive late to class will effectively be punished by a system that awards points based on the number of times a response is entered. This can be helpful in that it tends to compel students to arrive on time and be ready to go at the start of class. On the other hand, this will make some students unhappy, since parking problems and other factors may make it impossible for them to arrive on time each day. And, because with this system they will not get credit when they do not attend class, students will dislike this system because they must occasionally miss class for what they consider to be good reasons. In other words, because they will loose credit due to factors beyond their control, students object to what amounts to the attendance based rewards.

And then there is the problem of how to handle the times when students forget to bring their clickers to class. One option is to allow students to turn in a piece of paper on which they have indicated their responses to the clicker questions. This may be especially important when reinforcement is based on the number of correct responses, though it will add to the time required of the instructor. Another option is to tell students that not clicking in counts as an absence, regardless of the explanation. This approach can be justified, especially if you are checking for accuracy of answers, with an eye towards helping students who seem to need it. The turned in paper does not allow you to know how a student would have responded before hearing the answer, and so does not meet the expectations of the class, and the failure to click in means that the student did not contribute to basis of class discussion (when discussion is based on student responses).

Another strategy is to bring an extra clicker or two to class, if you have them, requiring that students using these clickers indicate their name so that they get credit for responding. The problem with this is that the clicker you provide the student will not be linked to their name, making it harder to determine if a student responded correctly or not. And, if students know that you have these extra clickers for them, they may be more prone to forget the one assigned to them. I find that two extra clickers are enough for most situations. The more I bring to class, the more people are likely to forget to bring theirs.

Rewarding how well a group does, rather than each individual, may reduce the pressure on individuals to perform well, reducing concerns about being singled out. However, the grouping option puts pressure on students to perform, which some will object to. Rewarding attendance, the number of times that group members click in and/or the accuracy with which they respond may produce the desired effect of getting students to participate and respond as accurately as possible. But, because some students in a group are likely to miss fewer classes or opportunities to click in, or answer more questions correctly than others, some will be advantaged by this system at the expense of others. In other words, students who take attendance and/or responding correctly seriously will be disadvantaged by those who do not, creating what can be a very disruptive undercurrent in the classroom.

In other words, someone is going to object, no matter how you reinforce the use of clickers. You need to find a system that works for you.

PART 2 | Using The System

NOW THAT YOU HAVE A BASIC UNDERSTANDING of the system, the next step is to consider how to use the system to achieve your goals. If you want to use the system to take attendance, to learn what students understand and to break up the monotony of note taking, your use of the system may be somewhat different from someone who has different goals. This section introduces the various reasons that an instructor might use clickers, and how to achieve those ends. It also describes some potential additional effects of using the system. While one may have specific goals in mind, instructors using the system to achieve those goals may experience other effects. Unfortunately, not all of the effects of using a classroom response system are positive; the hope is to prepare you for what might be unintended consequences of using the system. Importantly, once you are aware of the consequences of using specific techniques, you can determine if what you want to do with the classroom response system makes sense, and prepare yourself for the additional consequences.

This section also introduces the use of PowerPoint. In addition to the basics of using the program, included are some techniques that many users of this program are not familiar with. One need not use all of the power of this program to use clickers effectively, but knowing what the program can do for you may enable you to make better use of the system.

5 | A Basic Introduction To PowerPoint

PowerPoint is a computer program that allows the user to develop and present materials to an audience. Once created, materials from and for this system can be projected on a screen as a slide show, distributed in the form of handouts and/or presented on the World Wide Web. Because it is both a system for putting material into a sequential format for presentation and a system for presenting that information, PowerPoint is often used to present information to students in addition to being a means through which one can use clickers to gather information from students. The material can be fixed, like a standard slide that might be viewed through a slide projector, or fluid, the appearance changing as it is watched by the audience. Movie clips and sound bites can also be included in the presentation, making this computer program very useful for classroom presentations. In other words, PowerPoint is a powerful, multimedia tool that can be used to present static or dynamic information to students, depending on the ability of the instructor (user). Importantly, the basics of this system are easy to master; most people have little difficulty using the system once they make the decision to adopt it.

PowerPoint is an incredibly versatile product. Generally speaking, if you have access to the materials, you can probably find a way to get it into PowerPoint. If you don't have access, you can probably create desired materials relatively quickly and easily. One can change font, color, and background to create a virtually unlimited number of slides with the same content. In fact, in their effort to create a specific effect or appearance, some people get carried away by the versatility, forgetting to focus on why they are using the product in the first place. They may spend hours getting something to be absolutely perfect, rather than spending that time on other, more important, matters. If and when the interest in perfecting a slide or show distracts the user from why they are using the product, this potentially helpful product becomes more of a distraction and less useful.

From a slightly different perspective, PowerPoint is software produced by the Microsoft Corporation. In an effort to lead the market, this software is constantly being updated and revised. Its look and feel changes, to a greater or lesser degree, as new features are added and as new ways of doing things replace the old. Importantly, because Microsoft does not want to anger the millions of users of this software, slides created in older versions of the program typically can be used in updated versions without difficulty. This means that slides created for presentation in PowerPoint will work for years, unlike some other technology designed for the classroom. This is important because it takes time to create slides and a quality PowerPoint presentation. Once created, the same slides and other materials need not be recreated each time you decide to present them to a class. They can however be revised and edited as necessary.

As noted in Chapter 1, time is an important consideration when thinking about including new technology in the classroom. Everything you do takes time away from other things. This means that instructors need to think of the costs associated with including new technology even as they consider the potential benefits. The ability to reuse slides means that the time spent creating slides can pay off in later semesters, when they need not be recreated. This may mean that you want to spend more time when creating objects for presentation than you might have if you were only able to use it once, and as a result, a better product is produced. Further, if there is a slide that did not quite work as expected, instructors may be able to tweak or otherwise adjust the slide, rather than having to start from scratch.

PowerPoint is typically included as software when one purchases a computer that uses Windows and Microsoft Office. It can also be purchased as a stand-alone product, and is available for machines using other operating systems, including those produced by Apple Computer. That said, some of the software that makes the clickers work may not work on non-Windows software. If you do not use Windows as an operating system, you may need to purchase software that mimics the Windows system in order to use clickers in your classroom. Clickers will not work without this software, so make sure that this program is installed on whatever computer you intend to use to when using clickers

Because PowerPoint is so widely used, it is not surprising that there are several web pages that may provide helpful hints as to what the system is and how to use it. For example; http://www.actden.com/pp/ provides a good deal of information on the capabilities of this program, and why instructors may find it useful. Similarly, http://www.west.asu.edu/achristie/powerpoint/ provides a lot of do's and don'ts when using this software. Other guides exist on the web as well—you may want to do a search to find them. Also available is the help within PowerPoint (click on the help icon).

Furthermore, microsoft also makes tutorials available from their website (http://search.microsoft.com/search/results.aspx?st=b&na=88&View=en -us&qu=powerpoint+). That said, the basics of PowerPoint are easy to understand.

Using PowerPoint For The First Time

Figure 5.1 shows a blank PowerPoint slide (what a slide looks like before any information is added). Note that this image is what you are likely to see before you add any clicker software. This image also indicates the icons and commands that one might see at the top of the screen — the commands that one might use to modify a slide. The command line probably looks familiar - it is the basic set of Microsoft menus that are found in other programs, including Word and Excel. Similarly, some of the icons on the second line are also probably familiar to those used to Microsoft and other products. Clicking on the blank sheet of paper (far left), for example, allows the user to create a new document, whereas clicking on the opening folder means that the user wants to open an existing document, and clicking on the disk means that the user wants to save the document that they are looking at.

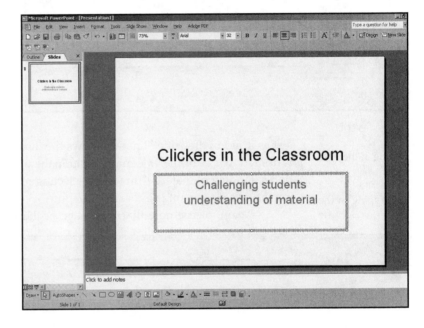

FIGURE 5.1
The Basic
PowerPoint Slide

Clicking within the box where it says "Click to add title" enables the user to write whatever text they want to be presented on a slide. For example, clicking within the box and typing "Clickers in the Classroom" changes the slide to reflect what was typed, indicated in Figure 5.2.

Notice that the image of the slide on the left also changes to reflect what is typed in the space for the title. As with other programs, using the icons B, I and U, in the second line of the commands at the top of the screen can be used to, respectively, **Bold,** *Italicize,* and underline the words as you type them. One can change the font, both style and size, by clicking on

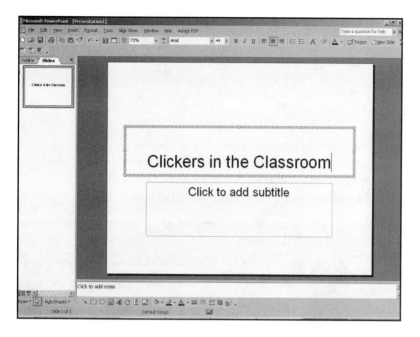

the appropriate commands. Similarly, one can change the color of the text, the alignment and so forth by selecting the appropriate icons. This can be done before you type the letters, or later (though, as in other applications, you must highlight the text you wish to change before you make those changes if you adjust these characteristics after the words have been typed).

Clicking in the subtitle region works

FIGURE 5.2
Typing in the Title

FIGURE 5.3
Typing in the
Subtitle

just as it did in the "title" line—it allows the instructor to add whatever text is desired. For example, one might include the words "Challenging students understanding of material" in red by selecting red as the color and then typing the text. The result is indicated in Figure 5.3.

Again, please note that the image of the slide on the left changes to reflect the changes made to the slide itself. It is possible to make more changes, add more boxes for example, or change other basic elements of the slide, but these you will learn later, as you become comfortable with PowerPoint. For now, it is important that you understand how to add material to a PowerPoint slide.

Another important thing to note here is that the slide is static; when presented it will

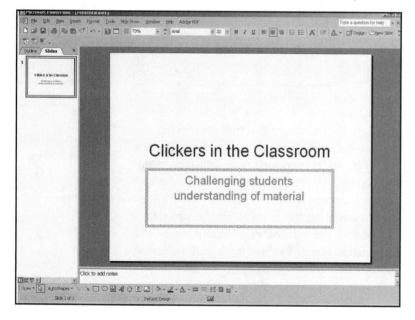

not change in response to students use of clickers or by attempts to adjust the appearance. As set up in this illustration, the slide will appear all at once, without, for example, any delay between the presentation of the title and subtitle areas as you may have seen done in other PowerPoint presentations. While adding such features is easy enough, and may eventually be desirable, it is unnecessary when learning the basics of PowerPoint, and so will be ignored here. When you are comfortable with the use of clickers, you should feel free to experiment with the various options that the program makes available.

Once slides have been created, it is easy to present them in a "slide show". Typing the F5 key, or clicking on "Slide show" on the top command line, then clicking on "view show" presents something similar to Figure 5.4. Note that the command lines and image of the slide that had been on the left side of the slide image disappear when you do a slide show. Other than that, the slide is identical to what was prepared. The image is larger, because there are fewer objects on the screen, but otherwise the image is identical to the slide as created.

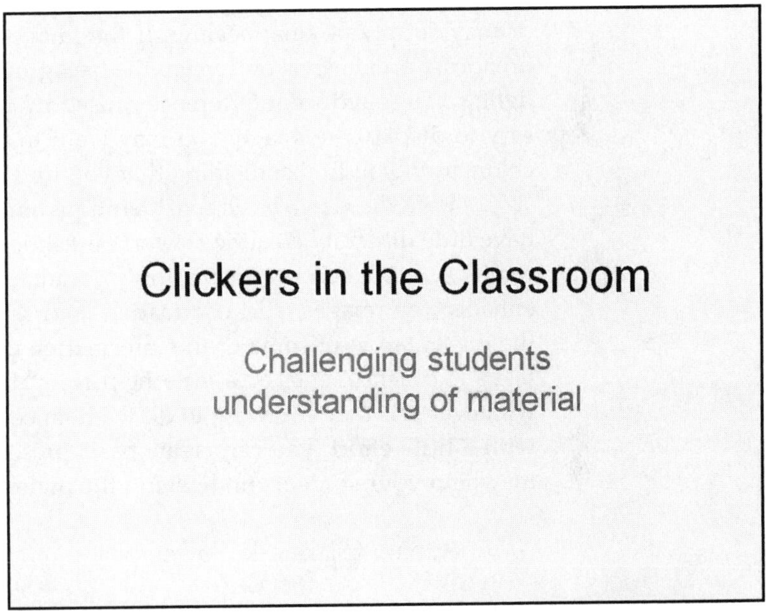

There are a couple of important considerations to keep in mind when preparing slides. First, make sure that the text is large enough to be read by everyone who will be viewing it. This means that you need to be sure that the font is rather large. How large the font should be depends on many factors, not the least of which is the size of the projected image. I suggest that, before you attempt to use PowerPoint in a classroom, you prepare a slide or two to play with, then present them in the classroom that you will be using. Stand at the back of the room, and in the corners, to make sure that the image can be easily seen by students sitting there. There are no specific font sizes or types that I can recommend. In general however, the larger the better.

Related to this is the amount of content that can appear on a single slide. The larger the font, the less material can be presented. Unfortunately,

Clickers in the Classroom

Challenging students
understanding of material

FIGURE 5.4
Slide Show of a
Basic Power-
Point Slide

some faculty forget this important fact as they prepare their slides. They may include more information than can fit in font large enough to be seen from the back of the room. The result is a slide that, though it contains a lot of information, is useless to those who are expected to read it. My advice is that, when you have too much material, present it on more than one slide. That way there is at least a chance that students will be able to read, and understand, the information you are providing to them. This is a situation where the less said on the slide the better. Use your voice to fill in details, using the slide for an outline rather than as the primary source of information.

A second consideration is contrast—the difference between the background and text. Too much contrast can be painful when students stare at the images for several minutes. Too little contrast may make it impossible to read the screen. Also, certain color combinations (red and green, blue and yellow for example) will not work for colorblind students or may simply be unappealing. If the images that you present are to be broadcast on a television screen, the background should be dark, the text light, so as to reduce interference caused by the amount of energy necessary to display the image. You may want to play with various fonts and colors until you find something that you find works well.

With these two basic considerations and a little practice, you should have little difficulty creating PowerPoint slides that work well in the classroom. As you become increasingly familiar with PowerPoint you can enhance the images, add contrasting colors, unique backgrounds, delay the presentation of some of the information on slides, and so forth. All it takes is practice, and/or a little help from those who are already doing whatever it is that you want to do to enhance your presentation. In effect, with a little effort, you can create basic presentations in PowerPoint that may help your students understand the material covered in the course.

Adding Clickers To PowerPoint

Clickers take advantage of the ability of PowerPoint to create and display dynamic images. The software provided with the clickers automates the conversion of a static image created in PowerPoint into a dynamic slide that reflects student responses. Because the software provided by H-ITT/InterWrite PRS/Qwisdom does the conversion, instructors do not need to learn how to make static images respond to how students answer the questions posed to them. In other words, the software that makes clickers work frees the instructor from having to learn how to convert static images to the ones that change within PowerPoint.

How to link your clicker system (H-ITT, InterWrite PRS or Qwisdom) is outlined in the materials sent with the software. You need not be especially facile with computers to install the software. I refer you to the help provided with your system to determine how best to do this. Further, I encourage you to play a bit with the various options available to you. While none may be perfect for what you want to do, both systems provide ways of displaying information that are quite good and probably very close to what you are trying to accomplish through the use of a classroom response system.

Regardless of the system you use, you will have the opportunity to use clicker slides that resemble Figure 5.5 as part of the slide show.

As students respond, the answers are recorded by the machine. Though the actual appearance will reflect the software used, this is indicated in Figure 5.6. (Note that this figure is based on the H-ITT Software).

Once data has been collected, it is easy to show what the results look like, again using the directions that come with the software. The result may be displayed as a graph, on top of the question, or in several other forms, depending on the software used. There are several options. I again encourage you to try various options to find the one(s) that meet your needs and those of your students.

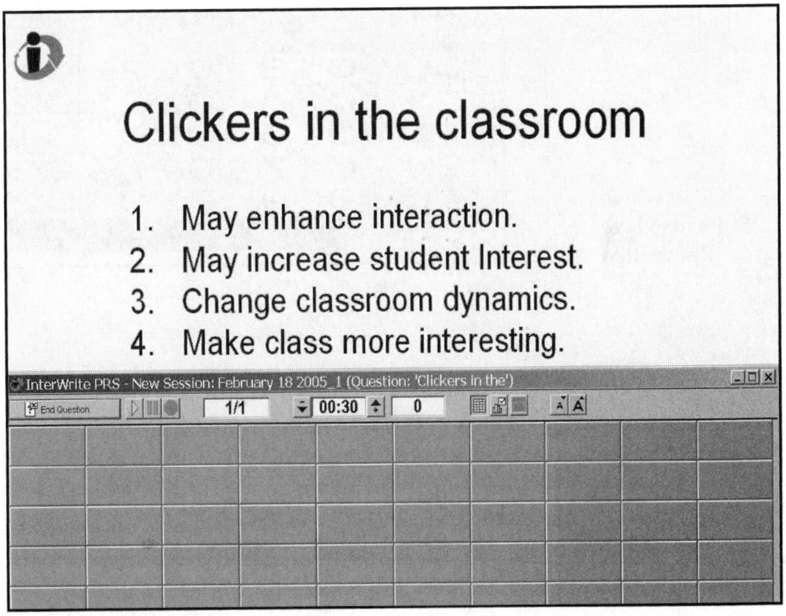

FIGURE 5.5
Slide Show of
Basic Four
Choice Slide

One additional feature available with the software is the ability to automatically grade responses. While this may make little sense for opinion questions (for example, How much do you like using clickers in the classroom?), it may make sense for others. For example, with a question like "In what year did Columbus depart on his maiden voyage of exploration?" there is a correct answer. You may wish to award students one or more points for answering it correctly. The program can do this for you, if you create the question appropriately. Again, you should check the manual that came with your software for instructions on how to do this.

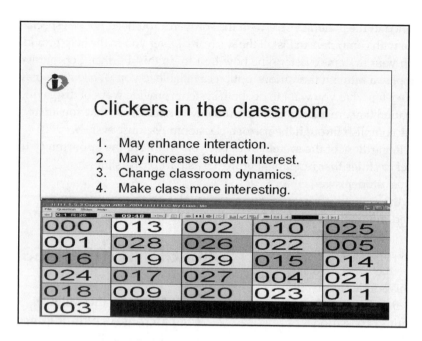

FIGURE 5.6
Slide to which
Students Have
Responded

FIGURE 5.7
The Effect of
Pressing the B
Key During a
Slide Show

Some Additional Hints

There are a couple of additional tricks that one may want to use when using clickers. For example, when a slide has been presented and it will be a while before a fresh slide will be used, consider pressing either the letter "B" or the letter "W"—for black and white respectively. This will cause the screen to go dark, or light, depending on which key you select. Figure 5.7 illustrates the effect of pressing the B key during a slide show. Note: this can be done in any PowerPoint presentation, not just those that involve the use of clickers.

To go on to the next slide when using either of these schemes,

either click on the letter again, or you can click the mouse to get back to the slide that was just presented (eliminate the black or white screen).

Similarly, moving the mouse around will produce a pointer on the screen. Use the pointer to highlight specific portions of the slide. Also, once the pointer is on the screen, clicking on the appropriate options allows one to adjust several features on the slide, as indicated by Figure 5.8.

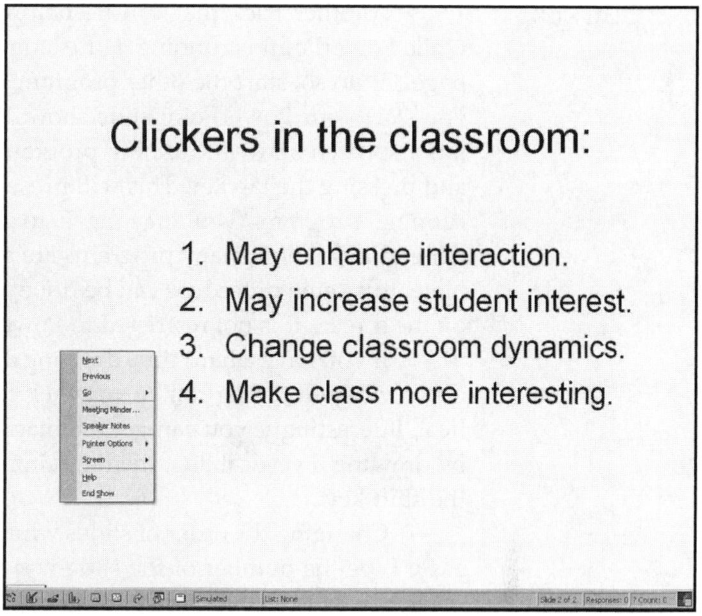

You can make the pointer into a pen (the difference between a pointer and pen is that the pen leaves a trace of where it has been, the pointer does not) by clicking on the pointer options, and selecting the pen option. You can also select the color of ink. You can also hide the pointer by electing this option from the pointers box. Figure 5.9 indicates the appearance of a slide after a pen has been used to highlight specific words.

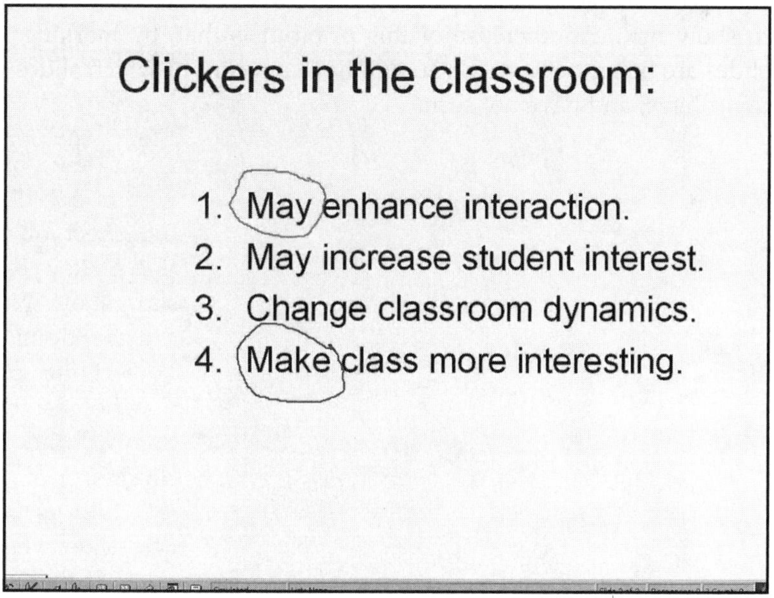

FIGURE 5.8
The Effect of a Right Click on a PowerPoint Show

FIGURE 5.9
The Effect of a Pen on a Power-Point Slide

Another trick that can be helpful is getting into other programs while PowerPoint is running. For example, you may want to link to a web page. To do so, start the other program, in this case a web browser, before you start your PowerPoint slide show. When in your PowerPoint presentation, switch between running programs by holding down the "Alt" key and pressing the tab key. This will present an image that indicates another running program (you may have to press the tab key several times, depending on how many programs are running, to find the one you want). Note, this same procedure can be done at any time to change the program on the screen, it is not restricted to PowerPoint.

If you have a hard time drawing a straight line, holding the shift key down while dragging can be very helpful. Lines can be horizontal or vertical. Interestingly, you can get the machine to draw a square or rectangle by drawing a vertical line, then moving horizontally, while holding down the shift key.

Changing the order of slides while presenting a PowerPoint show is easy. Type the number of the slide you want to present next, then hit the enter key. This enables you to skip slides, if appropriate, or to reorganize your questions based on how students respond to earlier questions. I find this very helpful when, in response to a question, I talk about something before I had intended to do so. Note however, this feature will not work while students are responding to a clicker slide. One helpful hint; make sure that you have the order of the slides available to you when you do this —perhaps a printout of the slide order that includes the number of each slide. This enables you to determine the number of the slide that you want to show next. Another use of this printout is that, by marking it while slides are being presented, you will have a record of which slides the students have, and have not seen.

6 | Taking Attendance

One of the primary reasons to incorporate clickers into a classroom may be that they can quickly and efficiently determine who is and is not in class. This may be especially useful in large enrollment classes, where it can take considerable time to call the roll. Once the system is set up to do this, taking roll is essentially automatic. Further, use of the classroom response system can indicate who arrived late, who left early, who understands the material, and who needs help. The only drawback, and it is a minor one, is that the linking of name and clicker number, which is necessary to use the system to do these things, is not automatic. It takes some time and effort on the part of the instructor or an assistant to make this happen.

Creating a class list may be as simple as cutting and pasting the names of students in the class into a spreadsheet used by the clicker program, but adding the number of the clicker used by each student can take a bit more time. As you might expect, the larger the class, the more time is required to get this system up and running, because someone must enter this information manually.

Students typically email their clicker number along with their name, or provide this information on a piece of paper. Someone (the instructor or an assistant) must then type this information into what amounts to a spreadsheet that indicates the name that is linked to each clicker used in the class. This takes a little time, and may involve a lot of e-mail. However, it must be done only once for each class; once the class list has the name of the students and their clicker number this information can be retained indefinitely.

As noted in Chapter 3, my students, and those of other instructors using a classroom response system, objected to the use of clickers as a way to take attendance. No one likes being required to be attend class, especially when they may have someplace else to be that could be more important, fun and/or exciting. However, if you, the instructor, believes that students miss something when they are not in class, invoking an attendance policy may make sense. This is especially true when you can be somewhat flexible—allow excused absences for example. Of course, allowing excused absences means that you or an assistant will have to listen to (or read) the explanations of why the various students did not attend class. Given that students may not always tell the truth as to why they missed class, it may be easier to excuse everyone, effectively negating any attendance policy, or to excuse no one, which then results in more student frustration.

A policy of treating people who do not click in as if they were not in class on a given day is going to result in some annoyance with the system. Instructors need to plan for this, and to develop a strategy to deal with it. There is no clear cut solution to this problem. That said, my solution is to bring a couple of extra clickers to each class, and have students sign them out if they need them. Because the number of these clickers is not assigned to an individual student, you will need to know who used which loaned clicker each day (thus the need for a sign out sheet).

Loaning clickers can create another problem however, in that students may forget to return the ones you loan out. What do you do if and when a student fails to return the one you loaned them? My solution is to threaten to fail students who do not return the clickers, but in a way that makes it clear that I am probably kidding. More appropriate might be to deduct points from students who keep the loaned clickers beyond the end of the class period.

Importantly, with the ability to take attendance comes the ability to determine how each individual responded to each question. Because the system links each clicker response with the user associated with that clicker, the instructor can tell if a student is answering correctly or not. While this may be helpful in identifying who needs assistance, it also limits the anonymity of responding. Unfortunately, anonymity is something that many students like about the system. This can be a significant consideration in classes where clickers are used to gather personal information, for example a human sexuality class where students are asked if they are gay, have had sexual intercourse recently, and so forth.

As such, if and when you use the system to take attendance, you may need to develop a policy for dealing with students who would rather you not know how they responded to each question. A policy of confidentiality makes sense as a way to assure students that their answers will not be discussed publicly. But you may want to go farther and assuring them that only the instructor will see their answers. If you adopt this approach, reasonable efforts must be taken to assure that no one else learns how a student responded. And, it may help to remind students that knowing how they responded is for their own good—it helps you, the instructor, determine who needs additional assistance.

Importantly, the use of clickers to take attendance is not fool proof. The system records which clickers are used. It does not know who is actually using them. This creates an opportunity for students to appear to be present when in fact it was someone else using their clicker. Further, it may be hard to spot who is using two or more clickers on a given day, especially in large-enrollment classes. To the extent that grading is based on clicker responses, there may be considerable incentive to have someone else click in for an individual student. Given incentive and opportunity, it may be likely that students will engage in this form of plagiarism. Thus, instructors using a classroom response system to take attendance need to develop a policy for dealing with students who click for their friends.

To Take Attendance Or Not To Take Attendance, That Is The Question

One of the joys of the clicker system is that students can respond anonymously. As such, you may get answers that are more honest if you do not use the system to take attendance. This is especially true when asking questions about personal behavior or attitudes, as indicated earlier. This is complicated by the fact that honest responses may be especially important in some situations, as when doing a course evaluation or otherwise asking for an opinion. This is especially true because students often conform to demand characteristics; they respond in ways they think the instructor wants them to respond instead of responding honestly. This bias in responding, often an attempt to win the favor of an instructor who has power over the student, can effectively eliminate the utility of clickers to accurately assess opinion. My solution has been to tell students that honesty is the best policy; that I am as likely to think highly of students who disagree with me as those who agree. This may reduce this presentation bias, but, in the end, it cannot assure honest responding to questions designed to explore opinions.

In contrast, there may be no incentive to respond in situations where responding is anonymous. Why bring a clicker to class, why respond, when you know that your response will not count or otherwise improve your grade and/or that your absence will not be missed. Because there may be a disincentive to respond (the student might be wrong), at least a few students will not bother to click in given any chance to do so. Also, the principles of operant conditioning suggest that behavior which is not rewarded will tend to drop out, so, with time, you would expect fewer students to participate in any class that does not have some reward for using clickers. In other words, unless they know that their responses are being noted by the professor, at least a few students will not take advantage of the classroom response system for feedback, maintaining attention, and so forth. For this reason, I suggest that you link names and responses.

Importantly, an atmosphere of trust may develop as the instructor earns a reputation of not disclosing how students respond to questions. While initially suspicious of my motives in asking questions like "are you gay", or "do you use drugs", my students quickly learned that the reason for asking such questions was to provide the class with information about the proportion of their peers who fall into these various categories. Because their responses were used as the topic of discussion (what percentage of people do you think are gay, drug users, etc.) and responses were used to answer the question rather than to explore who is and who is not gay or a drug user, students rapidly become accustomed to answering some rather personal questions in my classes. Further, most find the illustrations of the percentage of individuals who fall into these various categories interesting. More often, they find the actual percentages surprising. In effect, though I could determine how each student answered, students learned that I make it a point not to, especially for any questions related to opinion and/or personal behavior, note how any one individual responds.

Students typically can recognize the difference between questions used to assess understanding and questions used to illustrate a point. They appreciate that the instructor may find it important to determine who does not understand the material and who does, who needs help and who might be able to assist their peers. In fact, many appreciate the feedback, both positive (you understood what we talked about today), and negative (you incorrectly answered several questions today—do you need additional assistance on this topic). And they can appreciate a willingness on the part of the instructor to separate these two types of information. In fact, this may be a good way to explain your interest in finding out how each student answers, not as a way to grade, but as a way to acknowledge who needs help and who is doing well.

Another idea with respect to students who might rather keep some information from the instructor is to allow them to do so. My students are informed that they need not respond to every slide, though they are encouraged to do so. Given some of the questions that I ask, and that some students are very uncomfortable about self-disclosure, I allow students not to answer clicker slides that make them personally uncomfortable. Interestingly, over the course of the term, the response rates to the highly personal questions rise, as students learn that I don't care how they respond. In other words, appropriate use of private information may lead to increased trust, and increased self disclosure, in classes where this is appropriate.

In an ideal world, there would be no need to take the time to link students to the clickers they use. The system should report the name of the student along with their response. But the system does not work this way. Instead, the system records the student response along with an identifying number. That number is used by the system to distinguish individual responses. It can also be linked to the name of the student without much difficulty.

It would be nice if one could take a list directly from the registrars' office and automatically import it into the classroom response system. While this may be on the distant horizon, it cannot be done at this time. As such, perhaps the easiest solution is to copy an electronic version of a class list (either from the registrar, from a spread sheet, or from a word processed file into the clicker spreadsheet). This is easy enough to do using the standard copy and paste commands, and does not take much time. The more time consuming part of the process is in getting the students to tell you what their clicker number is, and entering those numbers into the spreadsheet.

The different systems (H-ITT, InterWrite PRS and Qwisdom) allow the creation of class lists, though what they call them and how they create lists are system specific. Given the number of different systems that Prentice Hall makes available, it does not make sense to describe how to do this here, especially where each company provides this information with the software. The important thing to remember is that you need to create the list only once each semester, and that the list can be edited if you wish, to reflect students who drop or add the course.

Once you have the class list entered in the participant list, you need to link the student name with the number of the clicker they received. This would be easy enough, but somewhat time consuming, if you handed the clickers to the students. In such a situation you could record the clicker number assigned to each student before handing them out. Unfortunately, however, because the devices are, in many cases, picked up at the bookstore or provided by someone other than the instructor, it will be necessary to find a way to get students to report the clicker number to you. I have passed around a sheet of paper on which students are to write the number found on the back of their device. This works well for the students who are present and remembered to bring their clickers to class that day. But some will be absent, and others will have forgotten their clicker. Then too there are occasional problems with interpreting the handwriting. I have also asked students to e-mail me, but some foul up the e-mail address, or otherwise cannot seem to send the necessary information to my account.

I strongly encourage that you set a deadline for getting the clicker number to the instructor, if only to stress the perceived importance of the classroom response system. Because the hardware costs money, students may be disinclined to purchase them, or to delay it as long as possible. Telling students that failure to provide a clicker number by the start of a specific class will result in a loss of points typically does the trick, and allows the instructor to hear from or about students experiencing financial hardship. I find that exceptions can then be made on a case-by-case basis, but a majority of those who might otherwise procrastinate are compelled to get their clickers in a timely manner.

Whichever strategy you use to gather the clicker number information, someone (you or an assistant) will have to type the clicker number in the appropriate column in the participant list. More correctly, you can either use the software provided by the manufacturer or you can type the information into a standard spreadsheet which is then imported into the system. I suggest that, especially for large enrollment classes, you use a standard spreadsheet and import the class as a whole to the classroom response system. This is because I find that entering each student, one at a time, requires more time. I import the class list to the spreadsheet (so that I do not have to type each name, and then type in the clicker number that each student has indicated).

You can expect to edit the participant list several times, as students remember to get the information to you and as they realize that they gave you incorrect information. Importantly, once you have the ID number for a student, you need not reenter it each time. The machine will recognize the link between student name and ID every time you run a slide show using that participant list.

I strongly encourage people to back up the participant list, to store it on a disk somewhere away from the computer you used to create the file. It would be embarrassing and time consuming to have to recreate the list (asking students to resend their clicker cannot help but make them laugh). And it is important to remember to bring this list with you when you carry your slides from your office machine to the computer in the classroom. If that file is not available to the machine where you present the slides, the responses will have to be recorded anonymously.

Using The Participant List To Take Attendance And More

Once the participant list that links the student name to their ID (clicker number) has been established, you can use the participant list to 1) determine whose clicker was used on that day, 2) how each individual responded, and 3) enable students to more easily recognize when their response was received. In other words, there can be significant advantages to the student and the instructor when names are linked to responses.

Importantly, to achieve any of these three goals, you must indicate the name of the class list before you start the slide show. All of the systems available effectively force you to select which class you are going to use during each session, they do not assume that you will always use the same class. This is because, using this strategy, it is possible to have some sessions where responses are anonymous while other classes link the student name with their response. While it might be nice if one could select specific questions to be anonymous or linked to names, this is not possible at this time. In other words, you can elect to, on one day, know how each student responded, whereas on another day you will not be able to determine this.

Select the appropriate list before starting the slide show and it will be possible to monitor individual student progress, record the number of times a student answered questions correctly, and take attendance. Note, you may have to browse your hard drive or the disk you carried to the classroom to find the name of the class list you entered, but once you have the list, you can use it in any and all class sessions throughout the semester.

Regardless of the software you use, the output from each class in these analyses is in the form of a spreadsheet. This means that you can save it in Excel or whatever spread sheet program you use, and, taking advantage of cut and paste, you can keep a record of how each student responded throughout the semester. Similarly, for those using such course administration packages as Blackboard and WebCT, this information can be uploaded directly into the grade sheets. This may be very helpful if you are grading based on attendance or the number of correct answers selected by each student—you can easily show students the classes in which their clickers were not used, and which questions they answered incorrectly.

Spreadsheets can also provide the instructor with information about how well individual students understand the answers to questions asked in class. For example, if one student answers a question very differently than others in the class, this will stand out when looking at how the class responded. This may indicate that personal contact with the student is appropriate—why was their response different from everyone else's? Similarly, where most but not all students answer a question correctly, you may want to contact those who answered incorrectly to find out if they now understand the material that they did not master initially. And one can do the reverse as well, noting who tends to answer most of the questions correctly, and send them a congratulatory note. This use of the information provided by the clickers tends to increase student interest in the course, if only because they know that you care. It may also tend to improve course evaluations, in part because you express interest in how students are doing.

Dealing With Student Anxiety/Distrust About Your Knowing Their Responses

Some students like the clickers because responses are anonymous. The fact that no one else knows how they responded to the questions, unless they tell them, has a strong appeal to those who are concerned about their image in the classroom or with their peers. Mistakes go undetected, which means that these students can seem to be better than they perhaps really are, at least in their own minds, if not yours. However, if you are linking students to their answers, some of these same students, and perhaps a few others may worry that their responses will reflect negatively on them. In other words, clickers may become a source of test/evaluation anxiety if and when instructors can determine how often students answer questions correctly. Not surprisingly, student anxiety related to how they are evaluated may explain why some instructors never link names and responses.

As you might suspect, clickers are most likely to create test anxiety or evaluation apprehension when students are graded based on the number of correct responses they make. In general, students will experience considerably less anxiety when grades are based on the number of class days on which a clicker is used than when grades reflect the number of correct responses. The anxiety produced by concerns about faculty, or peer, reaction to mistakes pales in comparison to concerns related to grading. Knowing that their grade may suffer if they answer a question incorrectly is enough to make at least a few students hate using clickers in the classroom. Further, it may increase the time it takes for some students to respond, as they ponder perceived nuances associated with each answer. Not surprisingly, grading based on the number of correct responses is a common objection to the use of clickers, as noted in Chapter 3.

This is not to say that instructors should not grade based on the number of correct responses. Instead, it is a warning to those who elect to use this technique; such use may produce a different student response than when grades are based on other criteria.

There are several ways to reduce "evaluation anxiety" associated with clicker use. For example, the instructor may base grades on the use of clickers (how many questions did the student respond to, or on the number of class days that the clicker was used), assuring students that you will not look at or base their grade on the number of correct answers. While some students will be worried initially that you may in fact be looking at how they respond, most who are anxious about this will become less concerned as the term progresses, and they see no impact of incorrect answers on their grades or interaction with the instructor. Eventually, most students will believe you when you say that you do not look at answers, an effect which will take less time over progressive semesters as students learn that you really do not look at how they respond.

Alternatively, you can make a game of it. Assigning students to a group, and rewarding group members based on the number of times or days that a clicker was used may reduce pressure on the students while assuring that they continue to participate. Because their "grade" or other incentive (you might give members of a winning group a pen, donut, or some other tangible but non-grade reward) is dependent on participation, and yet their individual responses are not the basis of the reward, this strategy may work well for students who are anxious about being evaluated. Further, it has the advantage of creating a bit of peer pressure, rather than from you alone. Because everyone in the group wants the reward, those who are not doing well are likely to get a bit of grief from others in their group, perhaps encouraging the poorer students to devote more time and energy to the class.

Of course this can backfire as well, as when one member of a group consistently misses class, or answers incorrectly and others give that individual grief as a result. Those who are contributing may take out their frustration on the individual who is not, resulting in anxiety, frustration, and ill will. In other words, while competition may spur people to excellence, it may also cause such adverse effects as anxiety, stress, and frustration. My advice is that, if you are inclined to use group, rather than individual performance, as the basis of a reward, try this strategy once or twice, just to see if it works for you and your class before you make it a part of the semester long system. Some people have had excellent results using this game approach, others find that it is a disaster.

This is not to say that you should not look to see if students are answering questions correctly. Even if you tell students that you do not check on their responses, it may pay to look at them every now and again. On occasions where all but a few students answer a question correctly, a personal and private note or comment from the instructor to the student expressing concern that the student is missing something, along with an offer to provide assistance works wonders. However, if you have indicated that you are not using the clickers to determine who is and who is not answering correctly, it may pay to find some excuse to talk to the student other than mentioning the clickers.

It should be noted that relying on clickers to determine who is and who is not in class may be somewhat flawed. Analysis of a clicker session indicates which clickers are and are not being used, not who is actually present. This is an important difference: students may bring a friend's clicker to class, along with their own. The result is that the instructor believes that two students were present when in fact only one was. Unless and until the only person who can make a clicker response is the person who purchased that clicker (perhaps based on the finger print of the user), there is always the opportunity for cheating. In a small class it might be possible to spot two clickers on the desk, or the student as they put one down and pick up another. But in a larger class, especially where seats or not tiered, detection may not be possible.

Importantly, infrared systems, because they do not record every response as soon as it is made, contribute to this type of cheating. If one click were always sufficient to record a response, it might be easier to tell when a person clicks in with more than one clicker; to see they made a response, put their clicker down, picked it up and responded a second time. However, because responses are not always recorded, it may be hard to tell if a student who put down their clicker and picked it up again is in fact using a second clicker. This is because the student may have picked the clicker up again when she or he realized that their response was not recorded. Alternatively, in picking the clicker up again, the student may be changing their answer. In other words, it is impossible for instructors to determine if a student is using more than one clicker simply because of the number of times they pick up a clicker and respond. Thus, though instructors need to be careful when they see someone apparently putting down their clicker and picking it up again, this is and of itself not good evidence that the student is using more than one clicker. Unless you see that a student has two clickers, it will be hard to determine if they are in fact cheating.

The hard part is actually catching students using more than one clicker in class. First, it is my experience that this rarely occurs, which means that I do not feel that I have to look for it. And, as the number of people in the room increases, the harder it is to spot someone doing it. That said, one system that may enable the instructor to determine when this occurs involves comparing the number of people who are in class on a given day with the number of people who clicked in. Obviously, you may have a problem when there is a larger number of clickers in use than students present that day. Unfortunately, this system does not tell you who cheated. Worse, it should be noted that the problem may exist even when the number of students present is the same as the number of clickers in use, since one or more students may have forgotten their clicker or may have been attempting to use a clicker with a dead battery. Another solution would be have to have a TA or the like monitor clicker use, in addition to the instructor. In this case, two or more sets of eyes may be better than one.

If you know that there is a problem, but do not know who is creating (which student is not present and which student is using two clickers) there is still the problem of dealing with it. How do you punish the persons responsible when you are not sure who they are? I know of one person who handles this situation by punishing the whole class—everyone looses points when this happens unless and until the culprits come forward. The advantage of this system is that it puts a lot of peer pressure on those who click in for themselves and someone else. But the disadvantage is that many students may be unfairly punished for the actions of one or two of their classmates. That said, this instructor tells me that she never has had a problem with people using more than one clicker.

It is my experience that telling students that the problem exists and that you are aware of it may be sufficient to end the practice, or stop it before it starts. Alternatively, you may ask that students report others who are using more than one clicker. I am not convinced that either of these strategies will, in the end, be effective. This is because both of these approaches make students aware that you are willing to treat the problem seriously. Another way to deal with this, one that requires a bit of staging, is to have an assistant pretend to click in for two students, and get caught. Confiscating both clickers, warning the assistant that they will face a disciplinary hearing and removing that person from the class has been known to be a very effective deterrent (until students catch on anyway). Note that this strategy will work well only for instructors with a flair for the dramatic.

Because many institutions have policies about it, one way to deal with students who click both their own and another's transponder is to warn students in advance that such conduct will be treated as plagiarism. Plagiarism is not just borrowing written words from another without attributing the source. It is passing off someone else's work as your own. The student who was not in class is effectively saying that they were in class when in fact they were not. And the student who clicked in for their friend is effectively condoning the practice of plagiarism by allowing someone else to plagiarize their efforts. Because there are policies in place for handling plagiarism, students understand the consequences of using someone else's clicker, or allowing someone else to click in for them, once it is made clear that this behavior falls under this stricture.

Another advantage of referring to the plagiarism idea is that most policies make it clear that others who know about this behavior are obligated to bring the actions of the individual doing it to the attention of the instructor. In other words, those who see their peers cheating can be held accountable if they do not report it. Making it clear that those around the plagiarist will also be held accountable may make it less likely that anyone will click in with a clicker belonging to someone else.

It should be noted that the problem of identifying who is in class is not unique to clickers. Unless you compare the name a student gives you with some type of photo ID, it is easy for one student to pretend to be another student. For example, unless you check the identification of each student entering the room, a fourth year student may be taking the test for a first semester freshman. Similarly, students have been known to use websites to obtain papers for their classes. Unless you watch them do the work, you never know whose work you are looking at. Students know that faculty do not generally check who is in the room, and have been known to do such things, much to the consternation of others.

7 | Improving Student Study Habits

Many instructors are concerned that their students do not read assigned materials in advance of class or otherwise come to class prepared. This may be because, with other responsibilities and things to do, students often put off studying until shortly before an exam. Or it may reflect a lack of interest on the part of the students on whatever the course they are taking is about. Clickers can be used to punish those who do not prepare for class and/or to reward those who are ready to learn.

Lack of responsibility, inability, or unwillingness to do homework is not characteristic of all students. In fact, a significant percentage of students in any class keep up with the work to the best of their ability. The real problem is that those that do not complete their assignments do not have the same level of understanding as those that have done the work. There is a tendency on the part of instructors to try to help those who do not get it, rather than working with those that understand the material assigned for any given day. Thus, the class tends to be slowed by those who did not do what was expected of them, less material is covered, and those that did the assigned work feel like the instructor is reading the book to them. Clickers can speed things along, by compelling those who might otherwise be unprepared to study before class.

Quizzing students about assigned reading materials and other homework assignments can motivate students who might otherwise not study to prepare for class. As you can well imagine, students are more likely to prepare for class when they know that they will be asked questions about the material than if no such questions are asked. This may reflect the possibility that their grade will suffer, that they may appear less knowledgeable than their peers, or that the instructor will think less highly of them if they do poorly on a quiz. Regardless of why, students who know that they are going to be quizzed are more likely to prepare for class than those who are not going to be asked about their homework. Thus, those instructors who want their students to come to class prepared may want to begin class with a quiz.

Importantly, grading may not be necessary to get students to do their work. Feedback to the effect that they do not know what they are supposed to, that they are unprepared, may be sufficient motivation to get many students to keep up with out of class assignments. That said however, to be effective, student responses cannot be anonymous for quizzing to be effective. If no one knows how an individual responds, there is little incentive for the less studious students to prepare for a daily quiz. Further, grading quizzes may increase student compliance with instructor expectations (Daniel and Broida, 2004) beyond that associated with ungraded quizzing. This is another way of saying that quizzes need not be graded to be effective, though grading them probably enhances the effect. Holding the student accountable for their performance is more effective than not. If students know that you will be looking at their progress, and can see that they did not know the answers to questions based on reading or other assignments, they may be more likely to prepare for class regardless of the effect on the grade.

That said, there are several potential problems with quizzing that may serve to dissuade faculty from using them to compel students to keep up with their work. Not the least of these problems is that the time required to administer a quiz is time that cannot be used for other activities. The time required to show the questions, and wait for responses, is time that might otherwise be spent lecturing, discussing, in small group activities or other forms of active learning. So, to the extent that using clickers is faster than using the traditional paper and pencil quizzes, the use of technology may be helpful.

Clicker quizzes take less time than those done on paper. First, there is no need to photocopy the quiz when using clickers (potentially saving considerable money). One copy, presented on the screen, is all that is necessary when using a classroom response system. Also, when using clickers, there is no time spent handing the pieces of paper to the instructor. In other words, clickers can save valuable classroom and department resources.

Another nice feature associated with using clicker quizzes, in contrast to those done on paper, is that students can compare their responses with others. For example, they may see that they are wrong, and that most of the class answered correctly. This may act as a disincentive to argue the point. Further, and more importantly, it may create considerable motivation to study more in the future. Or they may see that they answered correctly, along with the rest of the class, or that they alone figured out the correct answer. Thus, in addition to getting feedback as to the correct answer, with clickers students get a feel for how others in the class responded.

Another drawback of quizzing is the time that it takes to grade each student, and record the grades. This problem only gets worse as class size increases. Automated scoring takes care of a large part of the agony that many faculty feel at the prospect of grading quizzes. And the ease with which scores can be entered in a spreadsheet effectively deals with the concerns about recording. In other words, the use of a classroom response system can help faculty reduce the time required to grade and record quiz responses. This makes quizzing a bit more palatable, especially in large enrollment classes.

By including a few clicker questions each day that pertain to homework assignments, students will soon learn not to come to class unprepared. When clickers are linked to the students who use them, such questions can also tell the instructor who did not prepare for a class, something that at least a few students are likely to be concerned about. In effect, using clickers to determine if a student has done what is expected of them can be very motivating, helping students to do what they should do, which is to complete reading and other assignments outside of class.

Quizzing As An Effective Teaching Strategy

Many studies document the effects of quizzing on student preparation for class (Conner-Greene, 2000; Daniel and Broida, 2004). Students are more likely to prepare for class (read the text, do the problems, etc.) when such work is graded then they are for classes where no evaluation is likely. This same increase in effort occurs if students expect to be asked to demonstrate that understanding, as might occur with the use of a classroom response system. This is yet another situation where operant conditioning (reward and punishment) demonstrate their effectiveness in molding behavior. Instructors can take advantage of this, simply by asking a few questions each day, the answers to which come from whatever assignment was due that day.

Furthermore, quizzing has additional advantages that may help students remember course material long after the course is over. Daily and/or weekly quizzing serves to spread practice out through the entire semester. In other words, frequent quizzing makes it less likely that the only time students will study is just prior to an exam. Psychologists tell us that, in comparison to cramming for exams, consistent study patterns through the term are more likely to produce retention of information after the exam is over. While study throughout the term, rather than just before the test, does not assure that students will do better on exams, it tends to increase the ability to recall information long after a class is completed. Thus, many experts suggest that faculty use techniques like quizzing to encourage students to rely less on cramming and to spread their study time more evenly throughout the semester (Caple, 1996; Daniel & Broida, 2004).

Traditional quizzing is done with a paper and pencil; the instructor presents a few questions to which the student responds. The limited number of questions asked limits the amount of material that can be evaluated. The fewer the questions asked, the more likely it is that important topics or content will not be covered on the quiz. Thus, another problem with traditional quizzes is that they either take up a lot of class time, or do not assess a lot of information that the student should understand. Another reason to use clickers is that they take less time than methods do to find out what students know. And, for those who chose to use a classroom response system this way, the number of questions can be increased without reducing the amount of in-class time for other activities over what could be done with paper and pencil.

Another time saving feature associated with the use of clickers is the ability to count the number of times each student answers a question correctly. Developing the questions takes a little time. But grading and recording takes much more time, especially in large enrollment classes, and is very tedious, especially in classes with large enrollments. The machine will do in seconds what can take hours to do by hand. Further, these grades can be incorporated directly in to a spread sheet, course administration package (WebCT and Blackboard for example). To the extent that grading and recording are automated, using clickers reduces the time required in comparison to what would be necessary if one used paper and pencil quizzes.

The pop quiz is another option, adopted by some to take advantage of the effectiveness of quizzing while reducing the demands on the instructor. Because students do not know when they will be asked questions based on their assigned reading and/or other activities, they may be more likely to prepare for class than if they know that their preparation is not being monitored. Less effective than regular quizzes, because students hope that they will not be asked questions on a given day, pop quizzes are better than no quizzes, but produce more anxiety (Vogel and Collins, ND) and may produce lower grades. While it remains an option when using clickers, it makes little sense to use pop quizzes with this system, given the effect on student anxiety. Thus, pop quizzing may be a way to improve student learning for those who are too busy to develop regular quizzes. And, given the efficiencies associated with the classroom response system, those who use pop quizzes now may be more inclined to use regular quizzing to take full advantage of this learning tool.

Quizzing does more than punish students who did not prepare for class while rewarding those that did. Quizzing also enables the instructor to focus on the areas where students are having problems, rather than reviewing every concept covered in the text. For example, if students indicate that they understand a concept by selecting the correct answer to a clicker question, there is no need to spend time lecturing on that topic. Instead, when responses to a question suggest that a number of students do not understand a concept, the instructor can focus class time on that topic. Thus, the use of clickers enables faculty to know what students need help with, and what they already understand.

While I encourage the use of quizzes, it is not clear to me that quizzes need to be graded to be effective. For good students, knowing that they do not know, and knowing that the instructor knows that they do not know, is often sufficient to promote consistent study throughout the semester (distributed practice). Students who do not complete their homework assignments will find out that they do not understand the material as a result of the quizzing process. Knowing that they do not know, and that the material is likely to be covered on an exam, may be sufficient to get students to do what they should be doing anyway—reviewing the material more carefully. In other words, feedback from ungraded quizzes is likely to promote study, though perhaps not to the same level as that associated with grades.

It should be apparent that students will tend to dislike clickers when incorrect responses can result in punishment. Regardless of the weighting system students are going to experience some distress when the number of correct answers is used as part of a grading system. They will be more fearful of making incorrect responses when even a small portion of their grade hinges on how they respond to an individual question than if they know that how they respond is unimportant. And, if some questions are associated with more points, the anxiety level associated with answering clicker questions may become significant. When some questions have more weight or point value than others, students will worry that an incorrect answer on each and every question could significantly reduce their chances of earning an A. In part, this is because, unless you include that information on the slide, students will not know how many points a particular question is worth. Not knowing the answer may be bad enough, but the anxiety results from not knowing if they know the answer, and not knowing how much this lack of knowledge will hurt them makes the situation that much more aversive to the students.

The anxiety associated with grading clicker questions reflects what psychologists call test, or evaluation, anxiety. Grading answers makes the clickers part of the system whereby each student is being evaluated by the instructor. In contrast, answers that are not graded are more likely to be regarded as feedback, a learning tool. You, as the instructor, need to decide how you want to use the system (or individual questions), to grade students, or to tell them what they know and/or need to learn. When answers are not graded, students see the system as a way to help them learn. Grading changes the focus considerably.

That being said, it should be noted that there is no reason that you cannot elect to grade some questions, and not others, should you decide to do so. For example, ungraded opinion questions can be mixed with graded fact-based questions. It would not make sense to grade students on their opinions, but you may wish to grade the fact based items. Similarly, you may want to grade some, but not all fact based items. If you do this, I suggest that you tell students in advance of the presentation of a slide or use a key of some sort to indicate if that slide will be graded or not. For example, some instructors use color to do this; blue may indicate a graded question, black a question that will not be graded. (When using color, be aware that some students may be colorblind—careful consideration should be made as to which colors are selected).

Importantly, when using clickers to quiz students (for points or not) it is important to provide feedback as to why answers are correct and/or incorrect instead of just indicating the correct answer(s). While such feedback may not be important for those students who knew the answer, students who selected an incorrect answer, and those who were forced to guess will benefit from some discussion of the question after it has been answered. While feedback as to the correct answer is helpful, it is important for these students to learn why they were wrong, not just that they were wrong. Thus, I strongly encourage you to indicate topics, sections or pages that should be reviewed when an answer is incorrect. Or, if you have time, you may want to explain why each answer makes sense, or no sense at all. In other words, using the slide as the basis of discussion can facilitate student understanding of concepts on the slide.

Students will appreciate it when you take the time to go over slides that have correct answers. It is another indication that you care about their progress. You become less of an evaluator and more of a teacher when you can show people why they made their mistakes. Thus, when quizzing students, I present a question, get the class response, then discuss each of the selected answers, explaining why the one was correct and why the others incorrect. The idea is to encourage students to learn from their mistakes, not just to grade them based on how they responded.

Feedback After The Fact

Quizzes can be used to do more than simply help students study in a timely manner. They can also enable instructors to keep track of how well individuals are doing. If you find that someone is doing especially well, a quick e-mail praising their effort may go a long way. It increases the confidence the student experiences with the material, which may make them more prone to continue to study and enjoy what they learn. Furthermore, it may change the impression of the instructor, who then becomes more of a coach or teacher, less an evaluator or grader. As a result, your teacher evaluations may rise, even as your students learn more. In other words, a little positive feedback can do a lot for the student, and for you.

Unfortunately, not all students will do well. Some will need help. The hard part is often figuring out who needs assistance, and who does not. Quizzing can provide a hint as to who needs something more; those who consistently answer clicker questions incorrectly may benefit from additional attention. Thus, regular quizzing may give you a hint as to who is working, who is not, who needs help. And, when you find students who are consistently not responding to questions correctly, the instructor, or an assistant may be able to contact them individually, suggesting concern about the lack of progress by the student.

Telling students that you see that they are not doing well can also be an effective motivational tool, if it is done well. Expressing concern, without calling them stupid or lazy can tell them what they may not realize already, that they are doing poorly. Furthermore, it can enable them to access assistance that they need in order to succeed in the course. For example, asking students who are performing poorly if they would like help or benefit from a little time with you or a TA, or a quick note asking if something is wrong, may help students experience your concern about their progress without creating a negative reaction. Thus, combining quizzing with personal feedback, reaching out to students who are doing poorly on the quizzes specifically or in the class generally, may also do wonders for your teacher evaluations.

These steps do more than make you look better in the eyes of the students. They may also motivate those who are doing well to do better. They know you are watching and want to demonstrate that they are still worthy of praise. And those who are struggling may take more time to study, interact more with others in the class who are doing better, interact more with you to learn what they missed, and generally interact more with the material.

As an alternative or in addition to direct contract with students, once a class is over, some faculty post the clicker questions used in a class, and how the class responded to them. This allows students to see for themselves the questions that they answered incorrectly, those they understood and those they guessed correctly on. Posting also allows for additional time to compare their responses with how the class responded as a whole.

This additional information may help individuals understand how they are doing in the class, in comparison to others, and which areas that they need to study so as to do well on upcoming exams. Furthermore, when the instructor has gone over the material after presenting the slide, in reviewing the posted slides students may recall what was said about why an answer was correct or incorrect, or refer to an appropriate page to learn what they missed.

Posting has one additional advantage. When they know that the slides will be posted, students do not have to write down all of the information on the slide, reducing the stress in class, and allowing them to focus less on the answer and more on the concept. This may be the best reason to post questions and responses after they have been asked. Instructors can then spend less time waiting for the students to record the slide, and more time on the material.

Taking this posting idea one step further, some instructors post the questions for a short time (3-4 days), then take them down. This is presumed to force students to look at the slides shortly after they are presented. This, in turn, reinforces students who work throughout the semester, rather than just around the time when exams are given. Unfortunately, students have found a way around this one. They may get the notes on the slides from other students, or record what was on the slides, but not think about them until exam time. So, while posting slides after the fact may facilitate study, it does not necessarily spread the process of reviewing throughout the semester.

Examination By Clicker

Some faculty use clickers to administer exams. Slides of the exam questions are presented to the class, just as they might be during lecture. Students are given a fixed length of time to answer each slide, then the next question is posted. Feedback as to the correct response and how many students selected each response is not displayed. Obviously this will work only if student names and clicker numbers are linked; students will need to know that their response was recorded, and the instructor will need to know how each student answered each question. As a way of administering tests, this works well in some respects, but can also be disastrous.

The advantage of using clickers over a paper and pencil system is the immediate feedback that students receive. They can know immediately if they answered a question correctly or not, and so get a feel for how they are doing on the test. And, when allowed to do so, they can compare their answers with those of the rest of the class, giving them a feel for what they know in relation to others in the same class. Importantly, students can see how they do in relation to the class as a whole, but not in relation to the person sitting next to them or any other individual in the class.

Then too, exam grades can be calculated immediately and automatically, reducing the demands on the instructor. And, because they can be converted to a spreadsheet, grades can easily be transferred to course administration packages (WebCT, Blackboard, etc.). However, despite these advantages, I believe that exams should not be administered by clicker, especially in large enrollment classes.

The problem with administering exams with clickers is that students may, probably will, cheat. Specifically, they may use subtle or not so subtle cues as to what the correct answer is, broadcasting the answer to others who know the code. For example, different flicks of the wrist, selecting different receivers to send the information to (where more than one receiver is in the room), or simply whispering the answer have been used by students in my classes and others. The problem is that each student is seeing the same question at the same time. There is no question about which question is being answered. So, cheating is likely to be rampant when using clickers to administer exams.

Another problem is that the correct answer as indicated by the slide may be, in fact, incorrect. Most instructors have had the experience of the grading key being wrong. When the correct answer is presented to the student, and the correct answer is wrong, students who know the correct answer are going to be upset when the correct answer is marked incorrect. This will add to the stress experienced by the student, and may result in disruption within the classroom as students complain that they were robbed of points. This may explain why many who use clickers to administer exams do not present an indication of the correct answer once students have responded to the slide.

Then too, the testing environment does not allow for the additional feedback made possible by using clickers in the classroom. Specifically, unless the instructor takes the time to do so (an extend the anxiety of the students), there is no opportunity for the instructor to explain why each selected incorrect answer was wrong, no opportunity to explain why the correct answer is preferred, in an environment where questions are being posed continuously. As such, one of the primary advantages of the use of clickers, learning why answers are wrong, does not work well in a testing situation.

Entering Grades Into The Grade-Book

Once the slide show is over, the instructor can see how each student did that day. Further, the instructor can easily copy the information and then paste it into a grade book, text file, or spreadsheet on which the instructor is keeping tabs on student progress. Normally you will need to export the file from the program that runs the clickers, into a text or spreadsheet file before you can copy it directly into whatever format you need. While there are other ways to do this, highlighting the desired row(s) and/or column(s), then copying them puts that information into the "clipboard" is perhaps the easiest way to transcribe the information. As with other copy and paste or cut and past operations, going to the desired file, then pasting it there will enable you to use a single file to see how a student is doing. Note that you can add this information into either Blackboard or WebCT using this technique as well.

Importantly, this technique does not take much longer for 200 students than it does for 20. Thus, using clickers may speed the process of recording grades, making quizzing less time consuming for instructors of large enrollment classes. Thus, as noted earlier, quizzing, a technique which tends to compel students to study more than they might otherwise, can be made more effective and efficient by using clickers.

I find it helpful to keep a continuous record of how students respond. I create a spread sheet for the class, and copy the information about student responses onto this master sheet after each class. This enables me to spot trends (lots of correct responses, or lots of incorrect ones) and, at a glance, to indicate to students which classes they did not click in on (this is important because I grade attendance rather than the number of correct answers). While this can be done using the grade book function, I find that I prefer to keep a separate spreadsheet, just in case something goes wrong with the sheet kept by the program that runs the clickers.

8 | Promoting Active Learning And Student Interaction

Lectures can be very boring to those who are being lectured to. Some have suggested that students will tend to stop paying attention to the contents of the lecture long before the end of the typical class period. One report (Anderson & Ambruster, 1991) suggests that students record 50-70% of important ideas from lecture. Some people incorporate technology into their classrooms with this in mind; they are looking for ways to keep the students on task and attentive. Clearly there is much to gain if students can learn to pay closer attention to what is said in class.

Clickers can be used to break up the monotony of taking notes. In doing so they may serve to refocus attention. This is because students stop passively taking notes and start doing something else when they respond to questions using a classroom response system. Specifically, they must find the clicker, turn it on, figure out the answer to the question, select the appropriate response and check to be sure that their response was recorded. They may do more; learn from how others respond to the slide, question their understanding of the content, and stop sitting passively as the information flows to them. Because it is dis-habituating, repeating this process at odd intervals may help students pay attention in class, even when the clicker slides have nothing to do with the topic being discussed. Furthermore, because students are engaged, they are doing something other than taking notes, they are getting feedback on how well they understand the material being discussed, the use of clickers can change the passive lecture environment into one where learning is an active, engaging process. In other words, a classroom response system may change the passive lecture into something more interactive.

The incorporation of clickers may do more than simply break up the repetitiveness of taking notes. It often changes the nature of interaction within the classroom. Because students can see how the class as a whole responds, those who do not select the most popular or correct answer may be more willing to interact with those around them, asking those who answered differently or correctly, why they responded as they did. Similarly, students may ask questions of the instructor, either because they think they understand or know that they do not.

Use of the clickers may also change the way the instructor presents the course. Because she or he can determine when most people understand a concept, and when they do not, the instructor may adjust what is said to better meet the needs of the students. Because the instructor may need to move to the next topic or to elaborate a bit on what many students did not understand, the discussion becomes less predictable, and more interactive, based on what the students understand and what they need more help on.

Passive Learning Versus Active Learning

Active learning is a buzzword in education these days. Most educators have heard of it, many believe in it as a way to improve student learning, and more than a few college teachers have employed elements of active learning into their classrooms. Importantly, active learning does not require working in groups, demonstration of an understanding by completing a project, or developing an answer to an essay question. If the idea behind active learning is involving students with the material, making them think rather than simply react, and proving feedback about the adequacy of understanding, then clickers may be a way to create an active learning environment, even in large enrollment classes.

Unfortunately, many of the techniques typically associated with active learning do not work well in large enrollment classes. Few instructors want to be grading a hundred or more papers, evaluating that many projects and/or grading that many essay questions. Clickers may be an excellent solution for those interested in engaging students in their large classes, or smaller ones, in that the time required of the instructor is considerably less than it might be for other types of active learning exercises. And, because grading can be automated, it takes no more effort to use clickers in a class of 20 than it does in a class of 200.

Taking notes is a passive exercise. Few people can pay close attention to anything but the most engaging information for more than 20-30 minutes (Boreham, 1984; Johnstone & Percival, 1976). By interrupting the flow of information, by forcing the student to consider something, use of a classroom response system disengages the student from the listening and recording mode. It compels them to refocus their attention once the question has been answered. In other words, instead of just copying to paper what the instructor just said, using the clicker to respond to a question encourages direct interaction with the material as the students are asked to think for themselves. In effect, putting down the pen and picking up the clicker is a signal that it is time to engage the brain as more than an organ to store information. In the words of psychologists, use of clickers ends the habituation produced by the lecture environment, and promotes alternative activities, both of the body and the brain. The result may be the ability to refocus on the lecture once the response has been made, and an awareness that thinking is an important element when listening to a lecture.

Active learning involves reflection, consideration, and action based on that experience but it does not require putting something on paper. In fact, answering a clicker question may be evidence of active learning. Effective use of clickers disengages students from what they were otherwise doing, and encourages them to consider what they are studying. Clickers engage the individual student with the material, rather than allowing the simple recording of the information as it is presented. When students think, they engage. Good clicker questions force students to think, to actively engage with the material they have read and/or are hearing about for the first time in class. Using clickers to break up the recording of information, to increase the consideration of information, is an excellent way to bring active learning to almost any class.

That being said, it is important to note that the use of the technology alone will not assure active learning. The technology may encourage reflection, but only if the questions asked of the students promote this. Questions that ask for repetition of what was just said are not going to engage the learner as effectively as those that require the student to organize ideas, or to consider something beyond what was just said. Ideal questions are those that require students to think about the material, rather than regurgitate it. In other words, the use of clickers does not assure active learning, but using clickers in conjunction with questions that require thought, rather than memorization, will.

For example, one could ask a question about when Columbus set sail. Such a question requires that students remember the year in which his expedition began its voyage. It does not require that the student do much more than that. In contrast, asking about other events that occurred at about the same time may help students see the voyage of exploration in relation to other events that occurred at about the same time—the Spanish Inquisition for example. This could promote discussion not just of the discovery of the new world, but how that discovery was influenced by other events.

Most active learning strategies are not designed for classes with large numbers of students. They may work well for class sizes of less than thirty or so. But what can you do if you are teaching classes of 100 or more students? Small group exercises do not work well, because classrooms that hold that many students typically do not have chairs that easily can be rearranged to group students, and the noise generated by a large number of small groups may reduce the usefulness of group exercises. Reaction papers and quizzes take time, not only the time required to administer them, but also the time required to grade and record the results. The larger the class, the more time required to score and enter grades. Assigning, encouraging, and reviewing individual projects can be an onerous task, all the more so as class size increases. In effect, clickers may be the most efficient and effective active learning tool available to instructors of large enrollment classes.

When responding to thought provoking questions, students are no longer passive, They are engaged and actively involved with the material. Unless they are responding randomly (which can be discouraged by grading or otherwise looking at the number of correct responses), students are doing more than taking notes. They think about the material, and generate a response that reflects their understanding of it. Even better, feedback based on that response compels those who answered the question incorrectly or differently from a majority of other students in the class to consider the material from a different perspective. And the positive feedback that occurs when students answer a question correctly, or similarly to most others in the class, can also be helpful. In other words, asking use of clickers may provide a mechanism for everyone in the class to stop what they are doing, think about what they are learning, respond based on their understanding, and gather feedback on how consistent their answer was with respect to others in the class. The use of clickers changes the typical lecture class by breaking up the passivity that occurs when students only take notes. This is another way of saying that technology can make what had been a passive, boring, lecture, into an active learning exercise that provides important feedback to students about what they do and do not understand and how they are seeing things differently from others in the class.

Knowing that they may be asked a question about what they are writing down at some point in the immediate future, students may be more likely to think before putting pen to paper, as it were. Furthermore, clickers in conjunction with thought provoking questions allow (compel) students to reflect on what they have learned, and to determine the accuracy of their understanding of the material. Because students using this system are reflecting (developing an answer to the question), and self-monitoring (comparing answers with those of others and/or what the instructor indicates the correct answer to be), they are actively engaged with the material. The questions posed, and the feedback associated with the answer, act as a personalized message to the student, providing information about how her or his response compares to those of others in the class, and how the instructor views their response (correct or not). As such, use of clickers promotes active learning.

Importantly, clicking in and of itself is not evidence of active engagement with the material. Random guessing, responding without thinking about which answer is selected or without regard to what the correct answer might be, is not different from the passive learning that occurs during the traditional lecture. In other words, use of a clicker is not, in and of itself, evidence that the student is interacting with the material.

One way to prevent random responses from students who are not thinking about what the correct answer should be is to grade the answers to the questions. Many students will pay more attention to the questions, and think about what the correct answer is, if there is a consequence for responding correctly versus incorrectly. This explains the utility of grading responses to questions and/or letting students know when you notice that they are doing very well or very badly.

Similarly, knowing that they may be asked to indicate why they responded as they did may be a way to motivate students to interact with the material. Calling on individual students, and asking them to justify their response, will compel attention to the material and active consideration of the answer choices, but it may produce considerable anxiety. No one wants to be called on the carpet in front of their peers. Instructors should consider this possibility when using technique, as many students will dislike being compelled to defend their answers. This is likely to be reflected in the course evaluation, and an unwillingness to provide any answer on the part of some students. That being said, it is also likely to reduce random, thoughtless, responding if only because a lack of consideration may come back to haunt the student.

Importantly, the instructor need not call on any specific individual to justify their response to reduce the amount of random responding to clicker questions. For example, instead of selecting a name at random, the instructor can encourage active learning by asking people who selected a given answer to explain their choice. The important difference between this technique and calling on individual students is that, by not singling out students, no one feels forced to respond. This may reduce the anxiety associated with the use of clickers, and may increase the willingness to share ideas. This is especially true when students are not criticized for their choice of response. Comments from the instructor about how bad a response is are not likely to encourage participation. In contrast, statements to the effect that "I can see where you are coming from, but this is not a good answer because..." are more likely to promote class participation and active learning. Unfortunately, the strategy of calling on volunteers may promote random responding by those who will not volunteer, as they will never be called upon to justify their responses.

What do you do when no one volunteers? When I ask for people to defend their ideas, and no one volunteers, I try to come up with some possible explanation. For example, I might say, "if you were thinking that..., it might explain why some answered this way." This often draws students out who were earlier reticent to raise their hands earlier. If they see that their answer might make sense, more people are willing to admit to answering incorrectly. And, after a few non-evaluative comments about how this might explain why people answered incorrectly, students become more willing to respond when asked to justify their earlier answers. And, as more people become willing to volunteer, the number of people who do not think before clicking begins to drop.

One additional thing to consider when thinking of calling on students to explain their answer: the utility of this strategy goes down as class size increases. The more students in the class, the less likely it is that any one person will be called on. Students rapidly figure this out, which means that this may not prevent random responding in large enrollment courses. Then too, responding in large classes produces more anxiety than it does in small classes, making this a less effective strategy in large enrollment courses.

An alternative to singling out students and/or grading their responses is to keep track of how students are doing, providing occasional feedback when you find a student is not consistently answering correctly. You might note, for example, when entering responses in the grade book, that a student answered several questions incorrectly, and contact them about it. Knowing that the instructor can find out how they answered may motivate students to not respond randomly, simply because they do not want to hear from you about their poor performance in class. Knowing that they may hear from the "concerned" instructor about their poor performance may engage students who might otherwise not be overly concerned about how they answer each question. Similarly, one can contact those students who are consistently answering clicker questions correctly, praising them for their conscientiousness. As the word gets out that the instructor sends messages of praise, some students may be more careful in selecting their responses, hoping to get a positive evaluation from their instructor.

Another strategy to prevent random guessing, and to promote active learning, is to assign students to groups, and make answering correctly a competitive exercise. Making a game of it, getting groups to compete for points and/or some other minor reinforcement, may encourage students to think before they respond and to more actively engage in the class. Making responding a competition rather than a measure of understanding may change the focus of the clickers from an evaluation tool to a way to get something more out of the class. And peer pressure may encourage active learning on the part of people who might otherwise be disengaged from the class. But, as discussed earlier, conflicts between students may arise, as one group member consistently answers incorrectly and in doing so hurts the chances that the group will win whatever prize is available.

When using this game-like strategy it is important to consider how many students to put into each group. While there is no one best number, those thinking of using this technique need to consider that small groups may work well, because how each student does has more of an impact on how the group as a whole does. But, when few people are involved in each group, tension may result because some students may be frustrated by peers who fail to come to class, answer incorrectly even when trying, etc. Larger groups may prevent some of this frustration, because no one student has a great impact on the result. However, because one student does not influence the result very much, larger groups may produce less involvement on the part of some students.

Another way to promote active learning is to take advantage of group-think. Instead of having each student respond to questions, have one click in for a group. The idea behind this idea is to promote group process, discussion of different perspectives, and involve the students more with each other which occurs less often when students click in individually. This strategy takes much more time than having individuals respond for themselves, but may be valuable in situations where group process is important. Similarly, where a group comes up with an answer, it may be a good idea to ask each group to indicate why they responded as they did. Because no one individual is responsible for the answer, people are less afraid to respond to such questions.

Also, one can reduce random responding with clickers by including clicker or related questions on exams or other forms of evaluation. Knowing that the question they see in front of them in class may also appear on an evaluation will compel student attention in ways that other strategies may not. Wanting to do well on the evaluation, students who answer the question incorrectly will tend to pay close attention to what the correct answer is. In doing so they may learn why their answer was incorrect and another was a better choice. Furthermore, knowing that the question may come back to haunt them, those who guessed the correct answer may be more inclined to study the concepts addressed in the question a bit more than they might otherwise. And having an idea of what some of the exam questions may look like is likely to help everyone in the class feel a bit more comfortable when their responses are actually evaluated.

Note, you need not use any of these strategies for the entire semester. To the extent that students like variety, you may try one strategy for a day, week, or longer, then switch to another. As long as students know what you are doing, and perhaps why, they may actually enjoy the changes. Changing strategies are less likely to produce boredom, though it may produce some confusion from time to time. It is my experience that, consistent with my earlier observation that students like variability in the slides they see, changing strategies for preventing adaptation to the system can be very effective in maintaining student interest.

Increased Interaction As The Result Of Clicker Use

Clickers also promote interaction between students. Students have been observed to ask others how they responded, and, on occasion, why they responded as they did. This may be especially true when students are frustrated by the clicker slides (they answered one or more slides incorrectly). Other students may, wanting to look good (knowledgeable), or to be helpful, may explain the correct answer to those who misunderstood the question or otherwise answered incorrectly. Thus, there is incentive to ask, and to answer, questions related to clicker questions. The result can be helpful to all involved, though it may result in some disruption within the class as several conversations occur at once even as the instructor continues to direct the class in other areas.

Importantly, these conversations may occur over e-mail or within an electronic discussion board (where these systems are available to students), or verbally before, during or after class. Unfortunately, verbal discussion of clicker questions is most likely to occur immediately after the question has been asked. This may be seen as disruptive by the instructor and other students, especially when it occurs as the instructor or someone else in the class is discussing something else. Encouraging students to wait until class is over will not facilitate this type of active involvement with the material—students often have other things to do once class is over. My solution to this potential problem is to allow some discussion of slides after they have been presented. If students engage in private conversation at this time, it will be along the same lines as the class discussion, so not much is lost. Of course the instructor must then refocus the class when it is time to move on.

In talking with other students about the slides, students learn how people differ in their responses, and the factors that may have elicited one answer over another. In other words, they may become more aware of diversity of opinion. This is especially true when opinion questions

(Should marijuana be legalized?), in contrast to factual questions (____ was an early proponent of behaviorism), are asked. To the extent that appreciation of diversity is a desired goal in a class, asking more than a few opinion questions, and encouraging discussion of each answer, is likely to be an effective way to achieve that goal.

In effect, the use of clickers may help students become aware of the diversity of opinion within the classroom. As they see that others students select options that are different from the ones that they select, students may come to be aware that not everyone thinks the same way. This can be helpful in promoting dialogue about differences between individuals, and explains offers how background and experience influence understanding and an appreciation of individual difference. Depending on what a class is about, the instructor may want to use the diversity of responses to increase student awareness of diversity within the classroom.

To the extent that learning occurs from disagreement, discussion based on different answers to clicker slides can be a very positive encounter. And students may learn, rather then simply hear, why their response was correct or not. Hearing from peers may be more instructive than hearing from the instructor, since instructors often speak a different language or are otherwise seen as being somehow different from their students. Because students tend to see themselves as equal, and somewhat different from the instructor, hearing about it from another student may be more effective than hearing about something from the instructor. And, the interaction with peers will tend to be more active and involving than simply listening to what the instructor has to say.

Importantly, both students may benefit when a student teaches another. It has long been recognized that students who teach other students get at least as much out of the experience as the student being taught (Boller, 1999; Crouch & Mazur, 2001). Because she or he must prepare, organize their thoughts, and may review the material, the student teaching other student may learn as much or more than the person they are teaching. This widely recognized practice is the reason that many institutions create mentoring programs, teaching assistantships and the like. Thus, it may make sense to encourage peer to peer interaction resulting from clicker questions.

On the other hand, peer to peer interaction is not always useful. For example, some students may take advantage of being asked to help another student, either by the student or the instructor. Consider the possibility that the helpful student is actually providing disinformation to the person they are supposed to help, because lowering the one students grade may increase the changes of raising the helpful students grade. Alternatively, the helpful student may provide incorrect information because she or he does not really understand the question at hand. Either way, the result may be that the student requesting assistance receives the wrong information. Thus, if a student who is being helped by another continues to have difficulty coming up with the correct answer, a change in partner may be a good idea.

Clickers may also increase the amount of interaction students have with the instructor, in part because they may learn that they need help. Even as they turn to their peers, individual students may turn to the instructor when they realize that they are not answering questions as they should. This is especially true when students are graded on the number of correct answers they provide. But it may also occur in the absence of points. When a student realizes that other students are answering differently than they are, regardless of the impact it has on their grades, that individual may come to you with requests for assistance. In other words, feedback that they are not doing well may compel some students to seek assistance, something they would not do if they did not know that they did not understand what they were being asked to understand.

And, unless you take steps to prevent them, some students will chose to argue with you as to why their answer was more correct than yours. Students, especially those who are accustomed to being correct, will argue for their point of view when their answer is found to be incorrect. This is, of course, more likely to occur when the number of correct answers is reflected in a grade or when students are working in groups. If and when the instructor is comfortable with disagreement, allowing students to protest can be very helpful to the class as a whole, and for the instructor as well. Hearing their point of view may enable you to find where they are in error, or present you with a different perspective on the question. In other words, you may be able to provide feedback to the student about where they made a mistake, or they may be able to convince you that they were not wrong after all.

On the other hand, hearing that they are wrong may anger some students. They may feel that they know more about the topic than you do, that their hours of study more than adequately prepared them for whatever you asked about, and that your answer shows a lack of understanding. Dealing with such individuals can be difficult, to say the least. My approach is to show them as precisely as possible why I disagree with them, perhaps showing them a section of a text which supports my position. They may continue to argument, but at least they know why you say what you say.

Importantly, while clickers may be used to provide feedback, the questions asked and the answered provided may form the basis of class discussion. In other words, responses to clicker questions can lead to group discussions of answers and other active learning exercises. Unless the questions asked have absolute answers (what is the sum of 1 + 1), hearing disagreement can be the basis of other types of active learning exercises.

Interestingly, as students learn that discussion is ok, that they are not going to be punished for jumping in, more students may enter the fray. Relatively few students respond to questions from their instructor in the traditional lecture environment. When clickers are used, everyone has the opportunity to respond. Depending on the consequences of clicking in, more students may be willing to verbally respond to questions asked in class than in traditional classes, because they have gotten in the habit of responding. Some might call this the foot in the door technique - the idea that once students have started to respond to questions in the classroom, they are more likely to continue. On the other hand, it may be increased confidence resulting from answering clicker questions correctly that promotes their participation. While initial responses may be limited to the use of clickers, eventually you may find that people who otherwise say nothing in a lecture environment are willing to express their opinions. I find that about twice the number of students in classes where clickers are used are more likely to express themselves verbally than in classes that did not have access to the clickers.

Have The Students Write The Questions

Prentice Hall may have prepared a set of slides for your course. You can use them relatively easily. They take no time to develop, and very little time to organize into something that may be useful for your class. However, most if not all instructors using clickers in the classroom quickly realize that, no matter how good the slides provided by the publisher are, they can do better if they make the slides themselves. I find that most people who use clickers start by using the Prentice Hall materials, then, as they become more comfortable with the system, begin to prepare their own. The final step in this process is to have the students prepare them for you. Student prepared slides may not fit perfectly into your presentation, but they certainly require active learning on the part of the students preparing them. And students like the idea that they control events within the classroom.

Inviting, encouraging, or requiring students to develop questions to be asked during class sounds like a good idea. It actively involves them in the class, they learn about asking questions even as they learn the material well enough to ask a question about it, and they get a feel for how hard it is to develop good exam questions. As a result, they may be less critical of the questions they see on exams and quizzes, they may develop a better understanding of the material, and actively engage with the material. The downside however is that students take time to develop good questions, and some will not be able or willing to put in the effort necessary to come up with a good question. In other words, left to their own devices, student developed questions will not be as good as those you could write yourself.

While instructors may believe at the outset that getting the students to write questions will reduce their workload, this is simply not the case. Editing and encouraging students to fix the various flaws that will appear will take much longer than if the instructor were to develop the questions alone. In other words, you should expect that the use of student questions will increase, not reduce the amount of time that it takes to generate clicker questions. However, in some situations, the advantages in terms of active learning, writing, and the understanding that this technique fosters may be worth the extra time required of the instructor.

I strongly encourage that, if you take advantage of the opportunity to encourage students to write clicker questions, you work with students to develop good questions, rather than simply accepting whatever they produce. Several drafts may be necessary to create a useful question. In other words, you can expect that generating good questions from students will require more time than if you were to create them for yourself.

One thing that I like about having students generate questions is that, because I did not write the questions, I cannot be blamed for them. The student who wrote the question is responsible for both the grammar and content of that question. Holding them accountable can create a good deal of pressure on students to ask good questions, since they know that their peers will see what they wrote. The result can be that students put a lot of effort into this assignment, as they attempt to earn my approval along with that of their peers. Importantly, I do not tell the class who wrote a question, I leave it to the class to criticize the question without knowing who wrote it. This way people cannot pick on specific others, or are spared the brunt of negative opinion because they are not well liked personally.

Students in my classes are expected to develop questions that are grammatically correct, apply to the material being discussed, have at least one correct answer, and, in some cases, have answers that are not likely to sucker large numbers of students to answer incorrectly. For example, a student might want to ask the question "what is the sum of 2 + 3" in a psychology class, simply because everyone will answer the question correctly. However, unless I were lecturing on addition, this would not be either an appropriate topic for a question. Nor would it be hard enough to challenge students, something that I emphasize as being important when asking questions. As I tell my students, if everyone answers a question correctly, it does not discriminate those who know from those who don't, and so is not especially valuable.

Creating questions is an excellent exercise in critical thinking. Students and faculty must apply their understanding of the material. I encourage students to use contrasting terms when developing questions. For example, when writing a question about what a hypothesis is, I make it a point to include answers that describe a theory, an idea, and a postulate. The idea is to get students to use questions to compare concepts, not just to memorize what is in the text. Students report that they learn a lot by developing and answering questions for presentation in class, because they see nuances that they might otherwise have missed. In developing questions, students also come to recognize the difficulty faced by the instructor in developing exam questions, or using only the "good" questions from test banks.

Creating multiple choice questions where the question stem fits all of the answers can also be a challenge. Inexperienced writers often forget that each answer must fit with the stem, and it may take several revisions to get an acceptable question. Similarly, there may be problems with grammar, spelling, and a lack of understanding of the nuances of the various terms. Questions written by students, and those written by the instructor, need to be carefully checked to assure that they fit the expectations of the instructor, and of the students.

Even when the question is on topic, some students seem to believe that good questions are easy to answer. There is a tendency, at least on the part of some students, to ask questions that everyone will answer correctly. This will be especially true in classes where students are graded based on the number of clicker questions they answer correctly. Peer pressure, or the need of some members of the faculty to earn favorable faculty evaluations may explain why many people develop questions that everyone can answer. The idea is to assure that the question does not negatively impact how the students do in the class. Students need challenges, not just answers. Effective, thought provoking questions do not have obvious answers. Those skilled in testing tell us that such questions do not discriminate between those who know and those who do not, and so are not good questions.

Going one step further, questions that are so easy that everyone answers them correctly are not good clicker questions because one of the objects of asking questions is to promote discussion. If everyone answers the question correctly, there is nothing to discuss. Another problem with easy questions is that they do not require introspection, thought, or analysis. In other words, easy questions do not compel active involvement, and so do not promote active learning.

That said, humor can be a useful tool. Including humorous answers can be quite helpful in lightening the mood and providing some entertainment. And such answers tend to promote interaction between students, as they guffaw, giggle, and generally express their appreciation for the break in the routine. There is a danger however that too much humor may make it hard, if not impossible, for the instructor to keep the students on track.

9 | Clickers Change How You Spend Your Time

In considering the utility of clickers, it is important to determine if the costs associated with using this tool are greater than the benefits associated with their use. For example, if one is interested in using clickers only as a way to determine who is and who is not in class, one must weigh the financial costs of the clickers to the students and the time it takes to use clickers this way (in and out of class) in comparison to the increased number of students who are likely to be in class when attendance is taken this way. Similarly, if one is grading students based on the number of correct responses, it is important to consider the time required to develop and use the necessary slides in relation to how that time might otherwise be spent. And, if the goal of introducing clickers into a class is to increase interaction with and between students, one needs to consider the consequences of that increased interaction in terms of lesson plans and time for other activities. It is my experience that no benefit comes without a cost; one should not ignore the potential costs when thinking about adding something new.

As indicated in the previous chapter, using clickers in the classroom will change the dynamics of the class, both within and outside the classroom. Just as there are many potential benefits of using the system, there are also some predictable costs associated with incorporating the system. People unfamiliar with the technology need to think about both sides of this coin before they actually introduce clickers into their classes. This chapter explores some of the positive and negative consequences of using clickers in the classroom that people who are new to the system may not be aware of or otherwise think about. These include such things as less time to cover the material, more interaction with students, more time to prepare for class, more time to review student responses, and more and different interaction with students.

Using Clickers Reduces The Amount Of Time Available To Present Information In Class

People considering the incorporation of clickers into their classes need to know from the outset that using clickers takes time away from other in-class activities. In addition to the preparation time required to prepare the slides, the use of a classroom response system will reduce the time available for other activities because of the time it takes to show and discuss the slides. While this may be time well spent, the effect is to reduce the time available to introduce additional information, engage in other activity, and otherwise engage students. Thus, I encourage instructors to think about their use of clickers in relation to other things that they can and/or want to do in their classes, rather than viewing them in isolation.

If, like me, an instructor has a lot of information that he or she wants to convey during a limited class period, adding clickers to the classroom may make a bad situation worse. Time spent on clickers cannot be used to do other things. Thus, the time to do those other things is reduced when an instructor uses clickers in the classroom. For example, time spent discussing answers is time that cannot be spent introducing additional concepts or doing other in-class activities. Taking this into consideration from the outset can be helpful to instructors, who may limit their use of clickers and/or adjust how much material they want to cover in a given class.

Telling students in advance that you will not be able to cover as much as you might otherwise because you will be spending time with the clicker questions can be very helpful. It can serve as a warning that the students will be more responsible for the material than they are in other classes. You will not have time to review every topic covered in the text, for example, so they must master that material on their own. I believe that providing this information to students is also a good idea, as some may react negatively when confronted by a rushed instructor trying to do more than there is time for. Further, my advice is that instructors planning to use the system acknowledge this and make appropriate adjustments in what they intend to cover during each class period.

Knowing how much less to plan to present is problematic. There is no real way of knowing, short of personal experience, how much less time will be available for other things, once clickers are introduced. That said, I can tell you that I find that I can spend up to about a quarter of my in-class time involved with clickers —thirty seconds or so to present each slide, and, in a few cases, up to about five minutes discussing answers. Given the number of slides that I typically use, this means that I have time to present only 3/4 as much material as I did before I introduced clickers. Others report similar amounts of lost time, ranging from about one eighth to about one third, depending on the subject and instructor. Those new to the system should compare how much they cover when using clickers in comparison to what they are accustomed to getting through, to get some idea as to the costs of clicker use. This will help when planning future in-class activities.

Though I typically limit discussion to less than two minutes, there are times when this is impossible. When students do not recognize why they made a mistake, when they present alternative interpretations of the question and answers, or, having been prompted by something on the slide, students ask additional questions or share personal experience; slides can extend discussion and take time away from other topics. My solution is to cover less material than I might otherwise, requiring students to focus more on the reading assignments and other materials than they might otherwise. In other words, I tend to quiz students on their homework, in the hope that knowing that I am looking at how they do on the clicker questions will compel them to do it. This enables me to spend less time on the information covered there, and about the same amount of time as I would otherwise on information that is not available in the resources that students have available to them. This seems to work out well, especially because better students, those who would do the reading even if not quizzed over it, tend to do well on the clicker questions, embarrassing the others into doing what they should in terms of study.

In effect, I use clickers to compel students to keep up with the reading. Because they actually did the assignment, they are prepared for class. This means that I do not spend as much time introducing concepts covered in their daily assignments as I used to. I use clicker questions to reinforce what they should know, and then use the rest of my class time to introduce things that I would not expect them to understand or comprehend fully. Because students have prepared themselves for class, I find that I need less time to cover material in the book. In fact, this is where most of the time lost to clickers comes from in my classes. Because students have read the assignment, I do not have to tell them what they were supposed to know from the reading.

Another way to limit time lost to clickers is to severely limit or totally eliminate any discussion of material from the slides. In effect, using the slides as a (graded) quiz significantly reduces the time lost due to the use of clickers. Instructors may want to do this on occasion, if only because there is so much to cover on a specific day. However, as a general rule, this defeats one of the primary uses of clickers—providing feedback about what students need to focus on. An indication of the correct answer provides some feedback, but that information may not be sufficient to guide students as to where to turn for the answer, or why the response they chose was incorrect. To the extent that clickers are used in the classroom to promote learning, rather than as an evaluation tool, eliminating or severely limiting discussion is worth the class time required for it.

A third solution to this problem of time lost to clickers is to limit the number of clicker questions asked during each class. If and to the extent that the amount of time lost to clickers reflects the number of slides, then, obviously, one way to lose less time is to reduce the number of times that students are asked to respond. Depending on their purpose, instructors may not need to ask more than a couple of questions. Consider, for example, the possibility that one is using the clickers more to take attendance than for any other purpose. In this situation, getting responses at the start of the class (to compel students to be on time) and at the end (to make sure that they stay for the entire class) may be sufficient. In contrast, if the class lasts more than three hours, you may want to obtain responses throughout the session, to assure that students do not leave after the first slide and come back just before the end of the session slide is presented. Alternatively, if your goal in using a classroom response system is to help students pay attention, questions will need to be presented more frequently, making it harder to reduce the number of clicker slides presented in each class.

Unfortunately, no one has yet suggested an appropriate number of clicker slides per hour. It is not clear to me that there is one good number. The appropriate number of slides may vary by discipline, difficulty of the material being covered that day, the personality of the instructor and/or how much the instructor likes using the system. My experience is that, on a typical day, one slide every ten minutes (on average) or so works well for me. Others report using one slide every fifteen minutes on average, and some suggest using one every five minutes or so. To the extent that clickers refocus the attention of students, faculty who tend to be boring may want to use them more often.

Importantly, instructors can adjust the number of clicker slides used on any given day to reflect the content and how much other material they want to cover that day. Students do not seem to have an expectation of how many times they will be asked to respond, and altering the number on a daily basis does not bother or confuse them. For example, no one commented that I used four clicker slides one day, and 20 the next. This may, however, not be true in classes where students are graded based on the number of correct responses they make. When using this grading system, days where students can respond 20 times have more of an effect on a grade than those when there are only four opportunities to respond. Because of this, some instructors have been known to grade based on the percentage of correct responses on each day, rather than the total number of times a student answered correctly.

As discussed in the previous chapter, the use of clickers can dramatically change the amount of student to student and student to faculty interaction. There is a tendency to increase the amount of discussion, as students learn what others in the class think, as they discover what they do not understand, and as they develop a different relationship with the instructor. This can create real problems for the instructor. With increased interaction, the instructor becomes less in control of events in the classroom than when all they do is lecture. It may also create unanticipated opportunities for discussions of topics beyond the scope of the class, and for discussions about course related ideas that were not expected by the instructor. In other words, because of the tendency to interact more in classrooms where clickers are used, instructors may need to be a bit more flexible in how they handle the class.

Many students prefer not to get involved in classroom discussion. These individuals are also unlikely to answer a question posed by the instructor unless forced to do so. They elect not to raise their hands and may avert their gaze when the instructor asks a question, if only because they dread being wrong. The instructor will never know if these individuals developed a response because individual students who elect not to respond are rarely, if ever, asked to voice an opinion. This explains why the common tactic of asking questions (rhetorical or literal), and expecting students to answer them, may not be an effective strategy for promoting active learning.

As noted earlier, clickers change this reluctance to answer questions. Every student is expected to respond to questions, not just the few who elect to speak out. The typical result is increased willingness on the part of at least a few students to participate in discussion as noted in the previous chapter. In my case, about double the number of students participate in clicker classes than in classes when I do not use them. I am yet to find a way to involve everyone in discussion, but doubling the number is a significant improvement in my opinion. Thus, first time users should expect that there will be an increase in the number of students who respond to verbal questions, ask questions, and otherwise participate actively in the class. This increased interaction may also increase student willingness to discuss, argue or cajole points made by the instructor. The net effect is that the use of clickers is probably going to increase the amount, if not the level, of discussion.

Unfortunately, the increase in discussion may require more involvement of the instructor. To the extent that less confident people are more likely to speak up in a class where clickers are used, and to the extent that their lack of confidence reflects lack of understanding, there may be more incorrect comments by students when clickers are used. Thus, instead of being able to say something akin to, yes, good job, or correct, the instructor may find it necessary to correct the speaker more often in classes where clickers are used. I find myself correcting misconceptions more frequently in classes where clickers are required than in those where they are not, though I have not actually tested this hypothesis.

Because incorrect answers are perhaps more frequent when clickers are used, it is important that instructors find ways of correcting students who answer questions incorrectly without punishing them for speaking out. Students do not like it when they are talked down to, berated for answering incorrectly, and/or pointed out as being stupid. It should come as no surprise that responses that focus on whatever positive there is in the students response, rather than focusing on the incorrect elements that they expressed, is more likely to encourage others to speak up. Again, keeping the focus on teaching, rather than evaluation, may facilitate learning in contrast to creating additional anxiety in the classroom. Also, a side effect, one which may be very important, is that focusing on the positives may also tend to improve teacher evaluations.

In correcting erroneous beliefs, the instructor becomes less of a lecturer and more of a teacher. In other words, because the instructor knows that a student does not understand something, she or he can work to correct the misunderstanding, rather than moving on to other material. The effect on the classroom can be dynamic, especially as students who are less sure of themselves may ask the instructor to explain things again. The result may, again, be more time consuming than is typical when clickers are not used. More importantly, more students are likely to learn something from the discussion, because the instructor must correct the incorrect impression. In other words, instructors who use clickers may find themselves teaching more, and lecturing less than they did before clickers were introduced into the classroom.

Another possibility is the students actually understand things that the instructor assumes they do not. For example, I have often found that more students answer questions correctly than I assumed they would. This means that I do not need to cover that material as carefully as I might otherwise, sparing students from exploration of concepts that they already understand. The net effect is that I may spend a lot of time on some topics that I did not expect to belabor, and a lot less time on topics that I did. In other words, the ability to respond to students, rather than follow a planned sequence of events, is important when using clickers.

Of course, increased participation is more likely in classes where the instructor does not criticize students for this misinterpretation. Little discussion should be expected if students who answer questions incorrectly are yelled at or otherwise punished. If you want to be a lecturer rather than a teacher, yell, denounce, criticize or otherwise put down those students whose comments suggest that they do not understand the material. Otherwise, you should expect that the use of clickers will result in more questions, more incorrect answers, and more disruption of the lesson that was planned for the day.

I have observed that, on seeing that the response they selected was the one most often selected by the class, some students are more likely to speak up than they might be otherwise. Shy students may be more willing to present their views when they know that their ideas are consistent with most of the other people in the class than when they are unsure if they are in the majority or not. These individuals may believe that, because more people selected the response that they did than any of the other options, they chose correctly. This can create some interesting situations, as when the majority of the class responds incorrectly to a clicker question, since the instructor does not want to discourage the shy student from participating in the class.

Similarly, those who answered differently from the majority may feel a need to justify their position. They may mount a spirited defense, hoping to change the minds of the majority, and/or the instructor. These individuals, typically among the more confident students in the class, can be more difficult to tame than the shy ones who are less interested in participation. I find myself having to justify my position, teaching the logic behind answers, more often than I did before I used a classroom response system.

And, depending on how difficult the questions are, students may rapidly learn that the majority does not rule when it comes to correct answers. If this happens, shy students and those who distrust their understanding of the material are perhaps going to be less likely to speak up than they might be if the majority was correct more often. Thus, if your goal is to increase class participation, you might be advised to make most clicker questions so easy as to assure that the majority of responses are almost always correct. This however may conflict with other goals associated with the use of clickers, notably determining if students actually understand the material or not.

Importantly, students in classes where clickers are used are perhaps a bit more willing to interact with the instructor outside of class where there is an opportunity to do so. I found, and others have suggested, that there is more interaction between students and the instructor before and after classes that use clickers than in classes that do not use them.

However, this is not a universal phenomenon. It may reflect the instructor's personality, how personable she or he is, or perhaps how crazed that individual is just before or after class. It may be easy to discourage this increased willingness to interact with the instructor, by reacting negatively when approached by students, just as it may be possible to enhance this effect, by reacting in a more friendly and positive manner.

It Takes Longer To Prepare For Class When Using Clickers

Even though Prentice Hall provides slides for use with clickers, it is going to take a bit longer to prepare for classes that involve the use of clickers than it might to prepare for classes where this technology is not used. Selecting, arranging, and perhaps developing slides specific to your interest will take some time, as will adjusting the slides to fit your specific purposes. And involving students in the process of slide creation (see chapter 8) will add more time to this process. Importantly, it is how you use your time, not how much time you use that matters.

Almost everyone agrees that incorporation of technology into a class requires more time than it does if one does not bother to use this stuff. If nothing else, it takes time to learn to use, and then to become comfortable with, the equipment. Depending on what technology is used, it may be necessary to spend some time before each class setting up the equipment (if there is no one there to take care of this for you). And selecting, developing, and integrating materials into existing materials or new classroom activities also takes some time. Thus, you should expect that adding clickers to your classroom routine will require more time than if you were just planning to walk in and start lecturing.

It takes much less time to master the use of clickers than it does some other technology available for the classroom. Most people get the hang of clickers in an hour or less, whereas it may take four to eight hours to figure your way through such course administration packages as Blackboard and WebCT. Importantly, those who have never used PowerPoint are going to want to spend some time mastering that program in addition to the time required to learn the ins and outs of slide creation and other elements of clicker use. Fortunately, the process of mastering the software can and should be done well in advance of the first class. Many faculty spend portions of the winter or summer break becoming familiar with the software they plan to use, if only so as to be comfortable with it on the fist day of class.

Unfortunately, the additional time required to use technology does not end when you are comfortable with the system. Additional time is required throughout the term. Don't be fooled into thinking that using clickers will, at some point, require less time than not using them. It takes some time (not much, but some) to select which slides to use, arrange them, and to create those that are not already available, a process that will be required every time you use the technology. Similarly, keeping track of student responses requires some time, especially for those instructors who want to use the results of clicker sessions to determine who is and who is not getting it. Thus, you should plan on spending more time out of class when using a classroom response system then you did before.

One way to reduce some of the time and effort associated with the use of clickers is to have someone else, perhaps a teaching assistant, do at least some of the work for you. Given the information that goes on slides and the ease with which slides can be created, an undergraduate, graduate assistant or a secretary can modify and/or create the slides for you. Given the organization that you want to use, they can set up the PowerPoint session in the order that you prefer. And they can create lists of students who need help or perhaps should get a congratulatory note. Depending on your comfort with their doing so, this person may even send such notes to students for you. The idea is to reduce your workload by assigning the actual task to someone else. However, be aware that it may take some time to get that person to understand exactly what they should and should not do. Plan on taking time to review and perhaps revise the work the assistant does whenever you have an assistant do the work.

Another strategy to reduce the time required to use clickers is to use existing materials. Because it takes less time to use or modify existing materials then it does to create new ones, it makes sense to use what is out there whenever possible. Many people use slides prepared by Prentice Hall as the basis of the questions asked to their students. The publisher went to considerable effort (time and expense) to prepare slides in the hope that you would use them, thereby reducing the effort you must expend to use clickers. Though they probably did not create all of the slides you might want or need, their effort can be used as a guide. Further, they may provide a basis for the slides that you will create for yourself. My advice is that you use what you can, but not be afraid to modify what you have or to create new materials as necessary.

Not surprisingly, the slides created by Prentice Hall or anyone else for that matter, are rarely considered perfect by the instructors that use them. I, for one, have never met a series of slides that I could not improve on. Many others who use clickers apparently feel the same way. By the end of the semester, most faculty who incorporate clickers into their classes use few if any slides exactly as they were produced by Prentice Hall. Almost everyone has found something that they wanted to adjust, to improve in some small way or could be fixed in a matter of less than a minute. The necessary changes may be very minor, as in adjusting the order, adding color, adding sound or other options associated with the presentation. Alternatively, many people find it helpful to include a few slides of their own creation, perhaps to add more emphasis to points that they believe the folks at Prentice Hall underemphasized. Others develop slides that ask a question about something that they talk about but is not in the Prentice Hall content.

One factor that contributes to the need to adjust slides and to create new ones is the ease with which such things can be done. The software that comes with the clickers is so easy to use that many faculty, even those who swore that they never would, start modifying the content provided by Prentice Hall. Clearly, selecting the slides to use, and creating the ones that you feel are missing, is going to require some time, time that you would not have to invest if you were not using the clickers. It is just so easy to create the necessary materials.

Importantly, many instructors who have used these slides tell me that, in their first semester using clickers, they start using the slides prepared by Prentice Hall more or less exclusively. However, by the third or fourth week, many have created at least one of their own slides, and by the end of the semester, slides created by the publisher may be less than a quarter of the slides used. This is not because of the quality of the slides or the questions on them, but rather reflects the ease of use of the system, and an interest in creating a slide show that reflects the personality of the presenter.

Selecting, creating, and modifying slides is not a very time consuming process, I find that I typically create a slide from scratch in under four minutes. And, if I am struggling with a slide (it is taking longer than four minutes), I have been known to use it as it is. Slides are not limited to a set number of possible answers. Instructors need not spend a lot of time developing multiple alternatives to the correct answer, as long as they can come up with at least one. This reduces the time required to generate new slides, and explains why I have several slides with only two or three possible answers. Furthermore, as I think of additional alternatives, I add them to those questions that have what are, in my opinion, too few possible answers. Thus, questions that once had two or three possible answers now have six or seven!

I find that I spend more time going over my notes, thinking about where to put a slide, and what would go on it than I do actually creating the slide itself. And, in taking the time to do this, I review what I want to say, reorganize things, and generally update my notes. I have been teaching for a few years, and many of my lectures can now be delivered more or less by rote. One reason that I enjoy using clickers is that incorporating them into my classes has made me review my notes in more detail than I might have otherwise. In other words, some of the time spent preparing clicker slides is spent doing things that I should be doing anyway - refreshing my memory. Thus, not all of the time spent preparing to use clickers is wasted, some of it is spent doing things that I should be doing anyway.

In other words, preparing and modifying slides takes time but forces instructors to think about what they want to cover in the upcoming class. Going over the material may enable the instructor to refresh his or her memory about the material to be covered and/or freshen her or his approach to that information. Having taught the same courses literally dozens of times over the years, I find that I almost do not need to review my notes before going to class. In effect, spending time reviewing the questions that I will be asking, creating new ones, and modifying those that I used last time, helps me to re-familiarize myself with the information to be covered in lecture. I believe that the time spent working on the slides for the clickers also serves to change what would otherwise be a stale repetition of the things said over the years into something much fresher and more exciting.

Interestingly, less experienced teachers, covering a class for the first time, tell me that they like clickers because creating and organizing questions for the clickers helps them to become more comfortable with the material. In other words, taking the time to think about the upcoming lecture may not be an entirely bad thing.

Time Spent Reviewing Student Responses

Clickers do not automatically provide a list of names of the students who missed class. Nor can the software provide a list of how many times each student responded correctly without someone having to use several key strokes to obtain that information. Harder to note, but potentially useful is the possibility that there is a pattern in the responses of an individual student that indicates that that person gets it or is having a problem with the material. Similalry, Grades are not automatically put into a grade book. Some things you have to do for yourself.

Doing things like noting who was in class and who was not, figuring out if an individual student typically answers questions correctly or not, and combining information from the clickers with other information about student progress and performance takes time. This means that instructors who use the clickers to do such things are going to have less time to do other things than they would have had available if they opted not to use the system and/or gather this information. Regardless of how the clickers are used, instructors need to be sure that the benefits of using clickers outweigh the costs of using them.

Though the amount of time required to use clickers to take attendance is not great, a minute or so to generate a list of how each student responded to each slide on any given day, this is time that would be spent doing other things if one were not using clickers. And if one keeps such information, it may take another minute or more to add the information to a spread sheet that records how students responded throughout the semester. Importantly, the time required to note and record the number of correct answers is not different from the time required to note and record the number of times a student responds to a question, this is because the machine can do this scoring automatically. Because minutes add up, it is important to remember that it does take a bit more time to use clickers than to not use them.

Because it takes so little of the instructor's time, it may seem like a no-brainer. Use the system to take attendance and/or record the number of correct answers. Used this way, clickers will probably increase the number of students who show up to class each day, and compel them to study. In other words, spending a little time may reap real rewards in student performance. However, other factors must be considered when deciding whether or not to use clickers to take attendance or evaluate student understanding. One of the more important additional costs, at least for the students, is that of the anxiety that this may produce for some students (as discussed in Chapter 4). And, though it may be a cost or a benefit, instructors need also to consider the effects of clickers on class interaction as discussed in the previous section.

Teaching and/or administrative assistants may reduce the time it takes for the instructor to note who attended and who answered questions (in)correctly, but they will not result in the instructor having as much time for other projects as she or he would if clickers were not used. Why? Because it takes time to train the assistants, and assistants (like instructors) have been known to make mistakes. Because there is no foolproof way to assure that assistants are doing a good job, it will behoove instructors with assistants to monitor the performance of those individuals from time to time. Also, because the faculty member is removed from the data, at least to a degree, there may be an important cost associated with the use of such people. Faculty may miss information on how well they are conveying ideas and or helping students to grasp the material. This should be considered when determining who actually works with the data that comes from the clickers.

Importantly, though it takes little time to determine who is using clickers, it can take much more time to use this information to provide students with feedback on how they are doing in the class. Depending on what the instructor is looking for and how many questions are involved, it may take a while (several minutes) to determine which students, if any, are consistently not answering questions correctly. Also, there is the time required to contact the students who appear to need help, and the time required to actually help them (if they ask for it). Similarly, it takes time to spot and contract those who are consistently correct in their responses. The instructor must ask if the time required to do this is less or greater than the benefit the student receives from the feedback.

Faculty who do not use clickers or some other technology in their large enrollment classes typically do not know who understands the material and who does not until after a test has been given. Time spent developing this information is added to that time which an instructor would spend if she or he was not using technology. However, this may be a very productive use of time. Reaching out to students, giving them personal feedback on their progress in the course can be very helpful to the student, and result in improved teacher evaluations. To the extent that this makes what would otherwise be an impersonal interaction into something that is personal and helpful, I believe that this is time well spent. This is especially true in large enrollment classes—feedback from the instructor can personalize what is otherwise a very impersonal learning environment.

The important thing, when sending messages to students about their progress, is to not send too many. If everyone gets a congratulatory message, students will know it and devalue the message. The same holds true if the message says that the student needs to work harder, smarter or more effectively. Importantly, there are no standards for the percentage of students who should receive such notes. I find that, regardless of class size, I typically send notes to the five or so most accurate responders, and about the same number to those who need assistance. But what works for me in this regard may not work for others. This means that each instructor can determine what works for their personality, class size, and goals.

Clearly, the larger the number of notes sent, the more time required of the instructor or his/her surrogate. That said, those who send identical notes to students through a course management system like Blackboard or WebCT will find that it takes no more time to e-mail 50 students than it does to e-mail 5.

One factor to consider is if one wants to send identical notes to all students or if one wants to take the time to personalize the message a bit. Clearly, the latter option will require more time and effort than the former. The additional cost may not be worth the effort, depending on the circumstances of the class. Again, what you do and how you do it should make sense to you, even if others wouldHHating them into the class is unique and may use them to achieve different goals, there is no set of established best practices when it comes to the use of this technology.

Clickers Change How Students Interact With Instructors

Another thing to think about when using clickers for the first time is the possibility that the student-teacher interaction will change as the result of the use of this technology. Those students who normally are unresponsive in class will perhaps become more forthcoming as the result of being compelled to answer questions. They may respond more in class (as was discussed earlier), but they may also be more inclined to interact with the instructor before and after class (if such interaction is permitted or encouraged by the instructor). Students may be more inclined to ask for assistance, in part because they know that they do not understand something. They may also feel it appropriate to comment on the use of clickers, and other course related materials, perhaps in part because the old rules no longer seem to apply.

Students often think that their ideas or perceptions do not matter, especially in a large enrollment lecture class. Because they may never respond to a question or otherwise get actively involved with the class, they may come to think that they are there to learn, not to participate. Others may come to class to hear the truth by one who is all knowing, fearful that their ignorance and insecurities will be discovered. These beliefs may change as the result of being asked to respond to questions. The use of clickers may help students to realize that their ideas are not as far off as they worried they might be, that their ideas are consistent with others, and that they are often correct in their impression of the critical elements of the course they are taking. The result can be rather uplifting, and empowering.

The number of student e-mails I receive in clicker classes is about twenty percent more than in classes where I do not use the system. This may be because marginally "shy" or insecure students, emboldened by the experience with the clickers, may be more willing to approach the instructor with questions, concerns or ideas. Alternatively, it may be because students recognize that they are occasionally incorrect in their understanding that students are more willing to seek help from their teacher. Importantly, few e-mails addressed to me are complaints; most are legitimate questions about class procedure, answers to clicker questions, and/or questions that appear on exams or quizzes. The good news about such contacts is that they reflect student comfort in interacting with a member of the faculty. The bad news, of course, is that responding to them takes time away from other faculty responsibilities.

Similarly, knowing that the instructor is responsive, knowing that they are not always wrong, and knowing that there is a diversity of opinion, students may be more willing to approach their instructor before or after class, by phone, e-mail, and/or during chance encounters on or off campus. Furthermore, I find myself acting as an academic and personal advisor (I am a psychologist - this may be less common in other disciplines) to more students in clicker classes than I do for students who have taken non-clicker classes from me. Instructors need to plan for these "interruptions" in their daily routine, if only because they may become frequent (especially around the time when exams are given and when classes are scheduled for the next term).

Because clickers are a novelty on campus, and because I ask for feedback on their use, I often hear from students about what they think of this use of technology. Interestingly, I hear few complaints, with the exception that students ask why this technology is not used in more classes. In fact, every student I have asked has had something good to say about the system, though they may also find fault with it or how I use it. This may reflect a willingness on the part of students to say something that they believe I want to hear, a self presentation bias. However, I interpret this to mean that there is something about the use of clickers that students like, even if they dislike other elements.

For the most part, students are honest in their appraisal of the system. They are quick to point out flaws in the technology, and mistakes that I have made when using it. I appreciate their candor, and work to address any concerns they raise. The thing I find interesting is how many students actually take advantage of the opportunity to discuss this with me - at least forty percent of the students in each of the classes that I have taught using the system have had something to say about it beyond their responses to formal, in class, surveys.

10 | Using Clickers To Promote Active Learning

Left to their own devices, it is hard for students to pay close attention throughout the entire class period. Their difficulty in "staying awake" increases as the length of the class time increases, regardless of how engaging the class may be. One technique for dealing with this problem is to break the class up, have students stop doing what they are doing and start doing something else. Clickers can be effective tools in this regard. As students put down their pens or otherwise stop what they are doing, pick up their clickers, think about the question posed, determine their answer and respond accordingly, they become more alert and less dulled by the process of doing the same thing for an extended period of time.

But clickers can be used to do more than simply dishabituate students. They can do more than simply record and/or grade a response. They can be the basis of active discussion, promoting group process as students debate, argue, and cajole each other into adopting one position or another. Clickers can be used to promote student understanding in other ways as well, as when students are expected to develop at least a few of the questions seen by the class. Editing and revising student generated questions to make them more clear, more concise, and/or different in some respect adds another level of interaction with the material. In other words, clicker based interactions in the classroom need not always be about what the correct answer is, but can be used to develop more abstract understanding.

Using Clickers To Promote Discussion

Exam questions typically have a single correct answer. If the students selects that response, they get credit for it. But, because the clicker is used in a live, interactive, environment, there is no reason to have only one correct answer, or any correct answer for that matter, on the slides presented to the class! If the goal of the use of clickers is to engage students, then why not set the situation up in such a way as to virtually assure that students have to discuss? Ambiguous questions, questions that have no correct answer or multiple correct answers can lead to a discussion as to what the answer might be. These discussions can be held in small groups or may engage the class as a whole, depending on the needs of the instructor. The result is a very different type of interaction between students, and between the students and the material.

Consider the question "Which of the following is not one of the colors of the rainbow?" with the answers Red, Green, Blue, Indigo, and Violet. All of these colors are found in the rainbow, according to Sir Isaac Newton anyway, so the answer might be none of the above. However, if none of the above is not an option, students will have to select what might be an incorrect answer. They may select violet, since few people actually see that color. Or, if they think of the rainbow flag, they will be torn between indigo and violet since the color associated with these is purple.

Presenting a slide like this can be an opportunity for discussion, since there are different reasons to select different answers. With none correct, students are compelled to come up with the best answer. They can then be encouraged to defend their choice, in small groups or in front of the entire class. As an alternative, one could present a slide like this and then encourage discussion on an electronic discussion forum, as found in WebCT or Blackboard. The results of such interaction can be very enlightening.

Similarly, consider the question "President Bush is doing an excellent job" with the responses being limited to strongly agree, agree, disagree, and strongly disagree. (Including the option of neutral will enable many students to stay on the fence, and so may reduce the amount of discussion.) Unlike the question about the rainbow, this one specifically asks for an opinion - students will recognize that there is no correct answer. Most classes will have students who disagree with others with respect to this question. Getting them to explain their responses, to justify their beliefs can make a dull class that much more interesting. It will almost certainly energize the class, and disengage it from the passive process of simply taking notes.

Another way to promote discussion is to present statements that make no sense and ask students if they agree or not with them. For example, one might ask something like "The whether outside is cloudy" with the answers agree, disagree, and abstain. Clearly the word weather is misspelled in the answer—but some students will not recognize the problem. This can then be used as a basis of discussion about the importance of context, spelling, homophones, or whatever else fits the context of the class. Also, one can discuss the decision process —why did you chose to respond as you did given the nonsensible question. While this idea may be most appropriate for a psychology or linguistics class, it can also enliven a class in physics or mathematics, as colleagues in these disciplines discovered when they tried something similar.

On a related note, you may elect to misspell some words in a question or two, and encourage students who find the mistakes to contact you. This will promote interaction with students and may make the instructor less perfect in the eyes of her or his students. Some faculty award extra credit to students who find their (deliberate) mistakes on clicker slides, if only to reinforce paying attention to what is presented. To the extent that clickers are used to open up the classroom, to the extent that they are used to make students less fearful of contributing, little mistakes like this are useful.

Asking Related Questions To Promote More Integrated Thinking

Instructors need not think of clicker questions in isolation; combining questions can be an effective teaching strategy. For example, one way to help students understand the distinctions between concepts is to ask them a series of related questions. Thus, one might ask the question "Plato was a" with the possible answers nativist, empiricist, Roman, and physiologist. A second question might have the same answers, but ask "Aristotle was a"—the idea is to reinforce the difference between the concept nativist and empiricist, but also to provide a guide as to the difference between these people and those that followed them. One might follow these up with other questions about other nativists and empiricists, other philosophers and so forth, to reemphasize the concepts and help students understand the differences between concepts using multiple illustrations. The more different ideas are contrasted in clicker questions, the more students will come to understand the nuances of the differences between ideas.

Similarly, instructors can ask questions that get at the same topic from different perspectives. For example, having asked one or both of the questions mentioned above, an instructor might ask "Who of the following was most likely to do an experiment" with the possible answers Plato, Aristotle, Homer, Odysseus, and Theseus. Or one might ask which of those individuals was most likely to think of things in the ideal rather than the way they are. Again the idea is to approach a concept from multiple directions rather then a single perspective. Allowing students the opportunity to relate concepts and questions is likely to produce more integrated thinking then when questions are asked in isolation.

When asking multiple questions that are related, it is important to discuss each as they are presented, but also to note how the answer to the second question is related to the answer to the first, how the answer to the third is related to the previous two, and so forth, at least initially. The idea is to train students to think about the questions in relation to one another, not just to answer questions based on what they see on the screen at the moment. This may take some doing, as quiz and exam questions, the things that students are more familiar with, are typically unrelated. Many students will not see the relationship without being reminded of it. That said, depending on the level of the course, there are probably going to be at least a few students who relate questions and answers without assistance. Recognizing those students, and the ability to think about questions in relation to another, is a good way to promote this type of thinking.

Along these same lines, it is often helpful to encourage students to think about things that they just heard. Asking questions based on what was said a few minutes ago may help them recall, and put into perspective, information discussed more recently. For example, assume that you talked about the mean (average), then discussed the median. It might be appropriate to, having discussed both, ask about the mean before asking about median. The idea here is to make students think about, or focus on, things that are in less recent memory. In other words, because they do not find the answer in the most recently written sentences in their notes (or in what they just heard), students may review their notes and benefit from the process. In looking for the answer, they may actually refresh some of the ideas that they heard a short time before, helping them to retain that information. To the extent that review is helpful in retaining information, this strategy may promote retention and interaction with some ideas that might be on the way to being forgotten.

Have Students Write Or Revise Questions

Students have learned to answer questions by the time they get to your course. But do they know how to ask them? While many criticize the questions they find on exams and/or quizzes, few make suggestions on how to improve the question. Having students actually write questions for use with clickers may help them learn this very useful skill. Also, the feedback, provided by the instructor and perhaps their peers, may enable students (the person who wrote the question and others as well) to see nuances in the question that they may have missed originally, and has the advantage of allowing discussion of the question,

Having students write clicker questions has several advantages over asking them to write questions for exams or quizzes. Most obvious is the opportunity to get immediate feedback on their effort, not just from the instructor but also from their peers. To the extent that writing good questions is a skill that can be mastered, having students develop questions and having other critique their efforts can be a very helpful strategy for the student who wrote the question and others in the class. Suggestions on how to improve wording, how to increase difficulty, to offer other concepts that might be included in the question, and so forth can help students better understand the concepts involved. More importantly, such interaction is also likely to help them understand questions developed by others (as in the test bank for example). As they hear and think about improving questions they will learn how to answer questions in addition to learning how to ask them.

This strategy can be very effective in low to moderate enrollment classes (I suggest that you not do this in classes where there are more than about 40 students). If more students are involved, it is often too time consuming to have each student write a question and then spend time reviewing and critiquing it in class. A related strategy, one that works well in larger classes, is to present a clicker question and have students provide feedback on what they think the question is asking, how they think it might be improved and so forth. The difference between this and having students write the questions is that you or someone else wrote the question being evaluated. Thus, no egos are necessarily bruised. The opportunity to evaluate and criticize a question can engage students, regardless of class size.

A similar technique is to provide students with a question stem, and ask them to provide appropriate possible answers. This also provides an avenue for the discussion of test item construction, and a venue for the discussion of concepts relevant to the course. As students develop alternative answers, elements of grammar (do their answers fit with the question stem), elements of style (is one answer notably longer or shorter than the others), and content (does the question juxtapose two or more related concepts) should be noted.

A related strategy is to present a question and ask students how the question stem might be changed to make a different answer correct. Using the example noted earlier, when asking which of the following was most likely to do an experiment (Plato, Aristotle, Homer, Odysseus, or Theseus) you might then ask the class how to change the question stem to make Plato, or any of the other answers for that matter) correct. Alternatively, you can have students adjust the answers to make the question more difficult (pick a less famous philosopher, for example), or easy (have only one name associated with philosophy). This has the advantage of engaging the students, forcing them to remember who each of the individuals mentioned are, and to test their memory of the concepts linked to each answer. Note that these questions can be verbal, or they can be in the form of another clicker question. The point is to get the students to realize relationships between concepts, and this may take several slides to accomplish.

Clearly each of these approaches can be used with individuals or with teams. Depending on your goals, group processes may be more or less effective than having students work alone. One advantage of using groups is that this may reduce the amount of effort required of the instructor, if only because groups will produce fewer questions than will individuals. The disadvantage, of course, in using groups is that some students will rely on others in their group to do the work, resulting in unequal distribution of effort.

It should be noted that having students write questions is often more work than if you were to write them yourself. Depending on your willingness to edit, review rough drafts, and enter questions into PowerPoint at the last minute, the use of student questions may be so stressful as to make it impractical. However, there are real benefits to the students when this approach is used, which may make it worth the effort.

Use Clicker Questions On Exams

To further enhance the effectiveness of clickers, include a few of the questions asked in class on exams or other assessments. Because they want to do well on exams and other evaluations, students will tend to pay more attention to questions that have a chance of appearing on those assessments than those that do not. This will tend to increase attention to clicker questions, and make random responding less likely. But it will also tend to help students who answer incorrectly to focus on what they need to learn.

Putting clicker questions on the exam may also promote discussion of correct answers, as students who do not answer the question correctly look for assistance. Students in this situation are more likely to ask for assistance. They may do so in class, taking time from other activities, or outside of class, taking time from other things. Thus, though using the same questions on exams as appeared with clickers may increase attention in class, it may also have a negative consequence for the instructor, requiring more time than would be necessary if students remained unaware that they were not correct in their understanding of the material. Thus, those who consider using this technique as a way to increase interaction should consider the, what may seem to be, negative consequences in addition to the benefits. Depending on the personality of the instructor, and how willing he or she is to interact with students who ask questions, this can require considerable time on the part of the instructor, especially in large enrollment classes.

Importantly, students who answer questions incorrect and who need to know the correct answer because it will appear again on a graded assessment may also ask their peers for help. This is especially true when the instructor is otherwise unavailable. Thus, interaction between class-mates may increase when clicker questions are used on evaluations. Unfortunately, students are not always trustworthy when asked for help by their peers. In some classes, especially those that are highly competitive, students may be more interested in ruining a peers chances of doing well than in being helpful (providing correct information). There is also the pos-sibility that students will get help from people who are just as clueless as they are, which reduces their chances of receiving accurate information. Clearly the result may be increased frustration. Thus, care needs to be taken to assure honesty and accuracy when using this strategy.

In Summary

Because clicker use does not assure success, instructors need to con-sider the possibility that clickers will not improve the quality of their class-es. Because clickers can be used to achieve any of several goals, instructors need to consider what they want to accomplish in using this technology before they actually use it. Because some uses of the system prevent the possibility of reaching some goals while achieving others, instructors may need to consider which use of the system is likely to be most beneficial to their students. Because how clickers are used has such a large impact on their effectiveness, instructors need to consider how to use them to best achieve their goals. In other words, think before taking the leap into the use of this technology.

When using the system to present questions that have correct answers, I believe that feedback is critical when using a classroom response system. This feedback should be more about why an answer is correct or incorrect than it is about the response being (in)correct. Information about why is at least as important as information about how well each student did. Thus, taking the time to go over answers is at least as important as taking the time to present the questions and their answers.

When using the system to promote discussion, it is appropriate to ask questions that have more than one correct answer, and those that do not present any answer that is close to correct. Again, what is presented is less important than encouraging and guiding the discussion that follows. In other words, the use of the system is less important than how it is used.

Preparation and humor can make what might otherwise be a trying and frustrating experience much more enjoyable. When using clickers (and any other technology for that matter) it is important to anticipate that things will not always work as expected. Confidence and attitude can help students to feel better about the system, and about your use of it. And their trust will, in return, make your experience that much better. Thus, planning and preparation will do a lot to make the system work well for all involved.

And, finally, please do not hesitate to contact me, John Broida, Broida@usm.maine.edu, with any questions or stories about what happened when you (attempted) to use clickers to improve the quality of what you do. If you want to know more about something here, or have constructive criticism about the contents of this text, please feel free to contact me.

REFERENCES

Anderson, T. H. and Armbruster, B. B. (1991). The value of taking notes during lectures. In R. F. Flippo and D. C. Caverly (Eds.) Teaching reading and study strategies at the college level pp. 166-194). Newark, DE International Reading Associates

Boller, B. R. (1999). Non-Traditional Teaching Styles in Physics. Unpublished Opinion Paper, ERIC entry ED437111

Boreham, N. C. (1984). Personality factors related to self-reported lapse of attention during lectures. Psychological-Reports, 55, (1), 76-78

Caple, C. (1996) The Effects of Spaced Practice and Spaced Review on Recall and Retention Using Computer Assisted Instruction. Unpublished Doctoral Dissertation, ERIC entry ED427772

Conner-Greene, P. (2000). Assessing and Promoting Student Learning: Blurring the Line Between Teaching and Testing. Teaching of Psychology, 27, 84-88.

Crouch, C. H., & Mazur, E. (2001). Peer Instruction: Ten Years of Experience and Results. American Journal of Physics, 69 (9), 970-977.

Daniel, D. B., and Broida, J. P. (2004). Using web based quizzing to improve exam performance: Lessons learned. Teaching of Psychology. 31 (3), 207-208.

D'Inverino, R., Davis, H & White, S. (2004). Using a Personal Response System for Promoting Student Interaction. http://eprints.ecs.soton.ac.uk/9202/01/Using_a_personal_response_system_for_promoting_student_interaction.pdf

Duncan, D. (2005). Clickers in the Classroom: How to Enhance Science Teaching Using Classroom Response Systems. Boston, MA. USA: Pearson: Addison Wesley Benjamin Cummings

Elliott, C. (2003). Using a Personal Response System in Economics Teaching. http://www.economics.ltsn.ac.uk/iree/i1/elliott.htm.

Horowitz, H. (2002). Interactivity in a classroom environment. http://www.instruction.com/index.cfm?fuseaction=news.display&menu=news&content=showArticle&id=32

CLICKERS

Johnstone, A. H. & Percival, F. (1976). Attention Breaks in Lectures Education in Chemistry 13, (2), 49-50.

Montgomery, M. (2004). Personal Response Systems Enhance Learning. http://www.ucit.uc.edu/ucitnow/spring_04/sp04_prs.asp

Russell, J. (2003). On Campuses: Handhelds replacing raised hands. http://www.boston.com/news/nation/articles/2003/09/13/on_campu es_handhelds_replacing_raised_hands/

Vogel, H., and Collins, A. L. (N.D.)The Relationship Between Test Anxiety and Academic Performance http://clearinghouse.mwsc.edu/man scripts/333.asp

Wit, E. (2003). Who wants to be... The use of a personal response system in Statistics Teaching.